MANUFACTURERS AND MINERS

by
Irene M. Franck
and
David M. Brownstone

A Volume in the Work Throughout History Series

A Hudson Group Book

New York • Oxford

Manufacturers and Miners

Copyright © 1989
by Irene M. Franck and David M. Brownstone

All rights reserved. No part of this book may
be reproduced or utilized in any form or by any
means, electronic or mechanical, including
photocopying, recording or by any information
storage and retrieval systems, without permission
in writing from the Publisher.

Library of Congress Cataloging-in-Publication Data

Franck, Irene M.
  Manufacturers and miners / by Irene M. Franck and David M. Brownstone.

  p. cm. — (Work throughout history)
  Bibliography: p.
  Includes index.
  Summary: Explores the role throughout history of those who mine the land or work to make things, including factory workers, metalsmiths, mechanics, and vehicle makers.
  ISBN 0-8160-1447-7
  1. Labor and laboring classes—History—Juvenile literature.
2. Metal-workers—History—Juvenile literature. 3. Miners—History—
Juvenile literature. [1. Labor and laboring classes—History.
2. Miners—History. 3. Occupations—History.] I. Brownstone,
David M. II. Title. III. Series.
HD4902.5.F73 1988
331.7'62'0009—dc19

88-11694
CIP
AC

British CIP data available on request
Printed in the United States of America
10 9 8 7 6 5 4 3 2 1

Contents

Preface v

Introduction vii

Assayers 1
Factory Workers 5
Mechanics and Repairers 29
Metalsmiths 35
Miners and Quarriers 61
Power and Fuel Merchants 101
Vehicle Makers 113
Weapon Makers 125
Well Diggers and Drillers 145

Suggestions for Further Reading 153
Index 158

Titles in the *Work Throughout History* series

Artists and Artisans
Builders
Clothiers
Communicators
Financiers and Traders
Harvesters
Healers
Helpers and Aides
Leaders and Lawyers
Manufacturers and Miners
Performers and Players
Restaurateurs and Innkeepers
Scholars and Priests
Scientists and Technologists
Warriors and Adventurers

Preface

Manufacturers and Miners is a book in the multi-volume series, *Work Throughout History*. Work shapes the lives of all human beings; yet surprisingly little has been written about the history of the many fascinating and diverse types of occupations men and women pursue. The books in the *Work Throughout History* series explore humanity's most interesting, important, and influential occupations. They explain how and why these occupations came into being in the major cultures of the world, how they evolved over the centuries, especially with changing technology, and how society's view of each occupation has changed. Throughout we focus on what it was like to do a particular kind of work—for example, to be a farmer, glassblower, midwife, banker, building contractor, actor, astrologer, or weaver—in centuries past and right up to today.

Because many occupations have been closely related to one another, we have included at the end of each article references to other overlapping occupations. In preparing this series, we have drawn on a wide range of general works on social, economic, and occupational history, including many on everyday life throughout history. We consulted far too many wide-ranging works to list them all here; but at the end of each volume is a list of suggestions for further reading, should readers want to learn more about any of the occupations included in the volume.

Many researchers and writers worked on the preparation of this series. For *Manufacturers and Miners*, the primary researcher-writer was David G. Merrill; William Laird Siegel also wrote some sections. Our thanks go to them for their fine work; to our expert typists, Shirley Fenn, Nancy Fishelberg, and Mary Racette; to our most helpful editors at Facts On File, first Kate Kelly and then James Warren, and their assistants Claire Johnston and later Barbara Levine; to our excellent developmental editor, Vicki Tyler; and to our publisher, Edward Knappman, who first suggested the *Work Throughout History* series and has given us gracious support during the long years of its preparation.

We also express our special appreciation to the many librarians whose help has been indispensable in completing this work, especially to the incomparable staff of the Chappaqua Library—director Mark Hasskarl and former director Doris Lowenfels; the reference staff, including Mary Platt, Paula Peyraud, Terry Cullen, Martha Alcott, Carolyn Jones, and formerly Helen Barolini, Karen Baker, and Linda Goldstein; Jane McKean, Caroline Chojnowski, and formerly Marcia Van Fleet, and the whole circulation staff—and the many other librarians who, through the Interlibrary Loan Network, have provided us with the research tools so vital to our work.

Irene M. Franck
David M. Brownstone

Introduction

People who make things are the *manufacturers* of the world. The word *manufactured* originally meant "made by hand," for that is how most items were produced until the modern Industrial Age. Because manufacturing is such a widespread occupation, many craft specialties are discussed in other volumes of this series, especially *Artists and Artisans, Builders, Clothiers*, and *Restaurateurs and Innkeepers*. In this volume, we focus on metalworking, mining, and modern factory work.

Early humans made their tools and implements from naturally occurring materials such as stone, wood, or bone. But when, several thousand years ago, they first began to use metal, the craft of the *metalsmith* was born. Metal was far less brittle, and it could be worked in many different ways. In fact, metal was so important to early humans that archaeologists often classify their cultures by when they began to use certain metals—first copper, then stronger bronze, and then even-stronger iron.

For centuries, metalsmiths were generalists—that is, they tended to work with many different kinds of metals and to make many kinds of tools. Even so, some specialists developed: *Jewelers* worked with the most precious metals, especially gold and silver. *Locksmiths* and *clockmakers* did the intricately detailed work that remains their specialty today. (See *Artists and Artisans* for these highly skilled specialties.) Because gold and silver were so valuable, certain people began to specialize in testing the purity of precious metals. These were the *assayers*.

Some manufacturers specialized in making weapons. From early spears and bows and arrows, these *weapon makers* went on to develop the massive clubbing weapons of medieval times, the crude but effective guns and cannon of the Renaissance and early modern times, and today's truly powerful and terrible weapons of mass destruction.

The specialty of the *vehicle maker* is not as old, for weapons were known long before the wheel. But from about 5500 years ago, when wheeled vehicles such as the two-wheeled cart came into use in Asia, vehicle makers have played an important role in society. As people's lives came to depend on wheeled vehicles of all sorts, from farm carts to grand coaches, vehicle makers were found in every community.

For much of history, manufacturers tended to work on their own, with perhaps one or two helpers, and would therefore often make a whole piece themselves. An early gunsmith, for example, would actually make the whole gun, from the pouring of the molten metal to the assembling of the final pieces. But by late medieval and early Renaissance times, some manufacturers in Europe began to have other people, sometimes in large numbers, working for them. In these early factories—that is, "places for making things"—a rough kind of division of labor sometimes occurred, in which different parts of the manufacturing job would be done by different people. For example, one worker might build a coach body, another might build the wheels, another might assemble the parts, and another might paint the finished coach.

The lives of almost all manufacturers were transformed in modern times with the coming of the Industrial Age. For centuries, manufacturers had been skilled craftspeople, who turned out fine, custom-made items. But machines changed all that. Machines could turn out items one after the other—not as fine as handcrafted items, but cheaper and faster. Many manufacturers of custom-made goods could no longer sell their products, and they were either forced out of business or gradually opened factories themselves.

With the age of the machine came the age of the *factory worker*. Unlike craftspeople, who often had had years of training, factory workers with little or no training could operate many of the new machines. Division of labor became extreme, with each person being responsible for only one or a few tasks, to be repeated over and over, day after day. In the late 19th and early 20th centuries, machines in some factories became much more complicated. It takes considerable skill and strength to work on an assembly line in an automobile factory, for example. In the late 20th century, on the other hand, more and more work in some areas of manufacturing has come to be done by machines themselves, even by robots.

Supplying manufacturers with many of the raw materials they needed over the centuries have been the *miners* and *quarriers*. Whatever has been wanted—special kinds of stone, precious metals, or coal for heat energy—miners and quarriers have supplied. They have often done so at great personal risk, for their work has always been dangerous. When working underground, for example, miners always face the hazard of poisonous fumes and the danger of cave-ins, if the sides of the mines are not properly supported.

Power and fuel merchants have also provided much of the energy and some of the raw materials required by manufacturers. In earlier days they often traded in charcoal, made from wood, but today they more often supply energy from coal or oil. Oil itself, and always-necessary water, are often found by *well diggers* and *drillers*. Their work is also dirty and dangerous, but—like that of miners—essential.

In modern times, when so many products are manufactured at distant factories, the original maker is no longer available to fix them if they break down. So a new specialty has developed: *mechanics* and *repairers*. Sewing machines, automobiles, typewriters, televisions, toasters—all these and more are often fixed by the skilled mechanics and repairers found in every town and village. In this "disposable" age, however, some factory-made items are so cheap that, when broken, people find it easier to throw them away and simply buy new ones—a far cry from early times when every piece was painstakingly handmade and every scrap of metal was carefully saved and reused.

Assayers

Assaying—evaluating metals and minerals to determine their properties or market value—was not a specialized occupation until quite recently, but it may have been practiced as early as 200 B.C. Two 16th-century authors, Vannoccio Biringuiccio and Georgius Agricola, wrote about the importance of assaying in finding exactly where it might be profitable to begin mining operations, based on test samples. Assaying, then, was an integral part of mine engineering, but it was not an independent profession. Streetside *money changers* and early *bankers, goldsmiths*, and *exchangers* also dealt in the assaying of metals, especially the precious ones such as gold and silver. The people who were closest to professional assayers, though, were the *alchemists*. In their attempts to create precious metals by heating base (nonprecious) ores—minerals or mixtures of minerals from which metals can be extracted—these alchemists established the process of *dry* or *fire assaying*. They did this for their own research, of course, but also provided the service, for a fee, to customers.

The early *assayer* used essentially the same tools and techniques that the modern one does: balances, weights, crucibles, muffles, scorifiers, and cupels; fluxes, cupellation, reducing, oxidizing, and sulphurizing. Precious metals were assayed, as they are today, by crushing and washing a weighed ore. This was then combined with lead or fluxes (substances designed to ease the process) and melted in a small, shallow dish called a *cupel*. It was placed in the furnace's *muffle*—a kiln or part of the furnace protected from the direct flames. In the heating process, the precious metal was separated out from the unwanted refuse, called *scoria*. After the gold and silver were collected, the "button" of remaining lead was weighed to find out the amounts of gold or silver in the sample ore. Copper ore was roasted, crushed, washed, and blasted with heat. Lead-containing ore was crushed and mixed in a container called a *crucible* with borax (a mineral often used as a cleansing agent) and a glowing charcoal. Tin ore was roasted, crushed, washed, and finally mixed with two-thirds of its weight of borax; the mixture was placed in a hole in a stick of charcoal and then heated in a crucible. Iron was often assayed by *blacksmiths*, who burned, crushed, and washed the ore, and collected the metallic particles with a magnet. These particles were then heated in a crucible with saltpeter to produce pure iron.

By the 19th century, banks were offering assaying services to their clients, sometimes even commissioning specialized assayers. The United States set up assaying offices after the gold rushes in California and Alaska. Some of these "Wild West" assayers were apparently looking for more than a government salary, though. They took bribes to intentionally misinform and misguide gold *prospectors* about the value of their finds, allowing others to cash in on the claims. They were even known to hold

Assayers like these used age-old techniques for assessing the purity of metals. (Drawing by Bauer, from Georgius Agricola's De Re Metallica, *1556)*

onto precious stones they had labeled "worthless" and thus accumulate wealth for themselves. While most of the assayers were probably honest, there were enough dishonest ones to evoke violent actions from angry and frustrated prospectors.

From the late 19th century to the present day, assaying has become much more sophisticated. Advances in the science of chemistry have added *wet assaying* (chemical analysis) to the professional assayer's methodology. *Spectrographic analysis* has also become available, in which the wavelength of the light from a material is analyzed, using sophisticated machines to try to identify its chemical makeup. But many of the old traditional dry or fire assaying tools—furnaces, beakers, graduates (marked containers used for measuring), and crucibles—are still very much in use and preferred by many people in the field. Most governments maintain assaying offices to evaluate precious metals found within their country's borders as well as to determine the purity and value of those being passed through customs stations as taxable imports.

For related occupations in this volume, *Manufacturers and Miners*, see the following:
    Metalsmiths
    Miners and Quarriers

For related occupations in other volumes of the series, see the following:
in *Artists and Artisans*:
    Jewelers
in *Financiers and Traders*:
    Bankers and Financiers
in *Leaders and Lawyers*:
    Border Control Officials
in *Scientists and Technologists*:
    Chemists

Factory Workers

Factories are work establishments typically organized with highly specific division of labor for the purpose of mass-producing goods by using mechanical power on a large scale. Factory workers are often referred to as *blue-collar workers* to distinguish them from *white-collar workers* employed in clerical activities or business management.

Factory workers, in the strictest sense, have existed in large numbers only since the Industrial Revolution, which made such massive mechanical power available. But the evolution of the modern factory occupations can be traced back much further, to earlier forms of divisions of labor and standardized mass production. In this broader sense, *factory workers* have had a long history, even though large-scale mechanical power—and therefore true factories—have existed only for the last 200 or 300 years.

Most factories in ancient times were workshops or *manufactories*. These were small establishments that produced a relatively low volume of essentially handmade or handcrafted goods. In fact, the word *manufactory* originally meant "a place where things are made by hand." Some mechanical power was used, primarily the water-driven gristmills that ground flour for bread. Free *craftspeople* or *slaves* operated these workshops, generally with a minimum of helpers, slaves, or free laborers. Families often pooled their talents and efforts, making the hiring of additional people either insignificant or unnecessary. In such cases, the family business was handed down to heirs through generations, keeping the need for wage-labor at a minimum.

Most of the goods in the workshops were made with few or no mechanical aids other than simple hand tools. Products were not standardized, but individually and uniquely created, often to fill specific orders placed by customers. Only where a division of labor existed do we find forebears of the modern factory worker.

In the country, or in the small towns of classical Greece, Xenophon wrote in the fourth century B.C. that "the same workman makes chairs and doors and ploughs and tables, and often this same artisan builds houses, and even so he is thankful if he can only find employment enough to support him." Xenophon contrasted this situation with that in the cities:

> In large cities . . . one trade alone and very often even less than a whole trade, is enough to support a man; one man, for instance, makes shoes for men, and another for women, and there are places even where one man earns a living only by stitching shoes, another by cutting them out, another by sewing the uppers together, while there is another who performs none of these operations but only assembles parts.

This division of labor existed in many trades, such as metalworking, weaving, woodworking, and milling. But since most of these duties were performed by hand, craftspeople had little in common with the modern factory worker.

In Rome, labor was sharply divided among workers in the weapons and uniform manufactories, the only real mass-production industries of the ancient world. Free craftspeople were employed in these, as they had been in the larger Greek industries. Yet slaves were also used in large numbers for both skilled and unskilled jobs, and almost always for the dirtiest and most strenuous activities. Most of the large-scale production in the Roman Empire was carried on in the provinces or in the country, where cheap labor—mostly *farmers* and off-season *grape harvesters*—was readily available. The largest-known industrial complex of the Roman Empire was at Barbegal, near Arles in Provence, France. Its 16 waterwheels busily ground flour for about 80,000 people.

The waterwheels at Barbegal represented a tremendous savings in human labor and efficiency—factors that, centuries later, would inspire the Industrial Revolution. They had the capacity to produce 3,200 kilograms (28 tons) of flour in a 10-hour workday, compared to the mere 70 kilograms (154 pounds) that could be crushed by two people working a rotary hand-mill for the same amount of time. But Rome, like most societies before the 18th century, had an economy based on subsistence and barter rather than surplus and profit. Mechanization was not common, particularly in Rome itself, where emperors were more preoccupied with the problems of supplying employment and entertainment to keep the populace busy and satisfied. Mechanization was generally thought of as something that would unbalance society and create massive unemployment, with resulting social and political unrest. Suetonius once related how the first-century A.D. emperor Vespasian sought to protect Rome's free wage-earners by rejecting a laborsaving device:

> To a mechanical engineer who promised to transport some heavy columns to the Capitol at small expense, he

gave no mean reward for his invention but refused to make use of it, saying "you must let me feed my poor commons."

It was this kind of thinking that blocked any serious attempts toward industrialization for many centuries.

Between the time of the Romans and about the 12th century A.D., the European world moved away from even the modest levels of mechanization and division of labor that had been found in Rome. Elsewhere, many crafts did develop high degrees of specialization and divisions of labor, notably Byzantine and Islamic silk and linen production, and the Chinese manufacture of earthenware and fine porcelain. Through the Middle East and Spain such patterns of production came to Europe, first to France and Italy. Europe also had a renewed interest in mechanization after the 12th century. The result was that Europeans developed efficient and standardized forms of production that eventually would make the West the leader of the world's economy.

Perhaps the earliest "factory" workers were the women (both servants and free laborers) who were employed in workshops (*gynaecea*) to spin fiber and weave cloth. These workshops operated in the Byzantine Empire from the fourth century A.D. and monastic gynaecea operated in Europe as early as the ninth century, A.D. The Cistercian monks were probably the greatest medieval industrialists. They operated many factories, which eventually inspired the whole notion of mass production. In the 12th century they operated a water-powered triphammer forge at Fontenay, and by 1330 they owned between 8 and 13 iron-producing factories. During the same time they ran hundreds of large grain mills, which some consider the first factories in history. Large abbeys and several of the great estates developed into collection stations for raw materials, which might then be used to supply craftspeople and laborers in the workshops. The 12th-century Abbey of Corbie was typical of these protected workshop enclosures. It had three large rooms housing six *blacksmiths*, two *goldsmiths*, two *shoemakers, locksmiths, toolmakers* and *parchment makers*. Most of these workers were free laborers, clerics, or servants of some sort. Many of the estate workshops hired women as well as men, the former tending to the spinning and weaving, the latter to the metalworking, woolworking and leatherworking.

The great monasteries and estates often had a factory-like organization of their manufacturing craftspeople and laborers. Large stocks of tools and materials were supplied to the workers, who would use them to make finished goods. Many of the special iron tools used by *carpenters, masons*, and *shipbuilders*, for example, were provided by their employers. The laborers at these workshops would have had great trouble obtaining these costly materials, tools, and machinery on their own.

Workers were protected and provided for by their monastic or estate lords, and they often lived in workers' camps just outside the abbey, castle, or estate walls or boundaries. Within the "factory" itself, workers were divided into particular crafts, and sometimes into subdivisions within crafts. They were frequently supervised most closely by *stewards*, who managed the manufactories for absentee or preoccupied landlords and owners. These early factory workers were often peasants or women, who worked in such shops to add to the meager incomes they made from farming; others were craftspeople who were not wealthy enough to set up their own shops or did not have the stamina to endure the uncertainty of independent ventures. By and large, they were only a minor part of the labor force—and a somewhat maligned one at that. The men especially were criticized for doing humiliating work that was thought to be fit only for slaves and women; agriculture was considered a much more honorable undertaking in those times.

Standardization of production and growing mechanization increasingly made factories the means of manufacturing goods during the late Middle Ages and up into the mid-18th century. At that point the movement became so widespread and technologically advanced that it has been termed the *Industrial Revolution*. But, in truth, the "revolution" was really an "evolution," which progressed slowly but steadily in the three or four centuries preceding it. Likewise, factory workers did not arrive on the scene suddenly. Rather, their profession evolved slowly from ancient to modern times, until it suddenly became a significant factor in the economy and took on characteristics that sharply distinguished it from its historical foundations.

As feudalism began to break down after the 13th century, large, urban workshops began to arise. Notably, the revival of the ancient fast-wheel transformed potteries into huge industries turning out great numbers of pieces in the Rhineland, Stamford in England, and elsewhere. By the 15th century, these urban potteries were producing pieces with standard shapes and graded sizes. Meanwhile, monasteries were pioneering mechanical innovations in the tanning of leather, the fulling (shrinking and thickening) of cloth, and in metalworking industries. These were gradually picked up by a new class of bourgeoisie-capitalists, people who were seeking their fortunes in both country- and urban-based industries. They took full advantage of the new machines and techniques, to which they added the specialization of crafts and divisions of labor to further increase production. In a 15th-century water-powered textile mill in Flanders, for example, one piece of cloth underwent 26 different fulling processes.

Printing was one of the first completely mechanized industries. The printed sheet was almost totally standardized and mass produced. The first printing office was established in Mainz in 1447; 100 years later it had 100 employees.

In the 16th century, both Flemish and British woolen textile factories were moved to the countryside to avoid restrictive guild regulations. The factory workers went with them and camped out at the newly established production sites, creating the first factory towns that existed solely because of and for the factory and its workers.

Also important to the development of the factory occupations was the spread of clocks in Europe. Clockmaking was a handcraft that had been developed in China by the eighth century A.D., and in Europe by the 13th century. Clocks increased *production managers'* awareness of time, and consequently cost and efficiency. Work had to be faster, more regular, and with a miniumum of waste or deviation.

Craftspeople, including women, formed the earliest corps of factory workers. Urban guilds, created during the Middle Ages to protect and regulate crafts, were committed to favoring hand production over machine fabrication, and unique rather than standardized pieces. At one time, they also had supported total production by master craftspeople. But the specialization separating crafts and the divisions of labor within them took place in the craft guilds long before factories became important. In 1422 craftworkers in London were separated into 111 different guilds. The leatherworkers alone were divided into at least 13 specialties: *cofferers* (makers of decorative recessed panels), *cordwainers* (shoemakers—literally, workers in Cordovan leather), *curriers* (preparers of tanned hides), *girdlers, glovers, leather dyers, lorimers* (makers of metal pieces for harnesses and saddles), *malemakers* (makers of wallets or traveling bags), *pouchmakers, saddlers, skinners, tanners*, and *white tawyers* (bleachers of leather).

The division of labor and dependence on overworked journeymen became so great in the guild system that many masters had little to do with the goods they "produced"; some never even laid their hands on pieces that bore their insignia and carried their reputation. The guild craftspeople held tenaciously to a system based on restrictive rules and the notion that goods ought to be sold for a "just price," not a hefty profit. The just price supposedly insured the high quality of work, but in fact was used more to strictly exclude any competitors from the privilege of plying their trade within guild-protected territories. The notion of a just price was in keeping with the Christian ideal that people were born to a station in life, and they ought not try to improve their God-given lot lest society as a whole be thrown into social, political, and economic upheaval.

After the Reformation, capitalists consistently eroded the restrictive idea of a just price; the Commercial Revolution put the economy of Europe squarely on a money-and-profit basis; and the subsequent increase in trade created excessive demands for manufactured goods. The guild system was insufficient to keep up with these demands. Moreover, the guilds sought to curtail the only thing that could have helped them meet the new market demands—technological innovation. Craftworkers in Italy had resisted the use of highly developed silk-throwing machines in the 13th century and water-powered mills in the 14th, since these required only two or three *mill operators*, instead of hundreds of *hand-throwsters*. In the 15th century, Florentine guild regulations were passed to ensure that the warp threads (the vertical foundation of threads through which the weft threads are woven) in cloth were hand- rather than power-spun. In individual cases, of course, craftspeople themselves sought to lessen their burdens by adopting mechanical aids and power technology but, by and large, machines were seen as opposed to their vested interests.

As the guild system became increasingly incapable of handling new demands for more goods, capitalists arose and moved in to take their places. At first, capitalists simply supplied or "put out" raw materials and mechanical devices for home laborers (mostly women) to work into goods. This *putting-out* or *domestic* system strategically avoided guild restrictions by moving production into private cottages in the "unprotected" countryside. Sometimes such goods were even taken to city shops, where craftspeople were commissioned to add finishing touches. The textile and woolen industries, eventually the first to make extensive use of the factory system, were instrumental in popularizing the domestic system. The women and rural craftspeople who were employed in such a scheme were, for the most part, grossly exploited. To make ends meet, they worked long hours for little pay and no hope of promotion. They were paid strictly on the number of finished pieces they produced, which necessitated grueling workdays and tight scheduling. Yet they had no recourse; only the merchant-capitalists who sponsored their efforts had the funds to supply them with sufficient amounts of raw material or the means of marketing products for a decent profit. Since most domestic labor was carried on in separate, individual cottages, workers had little hope or even thought of organizing for their common benefit.

In industries in which many laborers were needed in one place in order to best pool their cooperative (though separate) efforts, semi-factory organizations were created under one roof. They had extensive divisions of labor, although most of it still represented handwork. The *needlemaker*'s shop pictured in Diderot's 18th-century *Encyclopedia* was typical of this arrangement, and Adam Smith gave us an illuminating description of a pin factory from the period:

> One man draws out the wire, another straightens it; a third cuts it; a fourth points it; a fifth grinds it at the top for receiving the head; to make the head requires two or

Division of labor, as here in a needlemaker's shop, is a practice with its roots in Classical times. (From Diderot's Encyclopedia, *late 18th century)*

three distinct operations; to put it on is a peculiar business; to whiten the pin is another; it is even a grade by itself to put them in the paper; and the important business of making a pin is in this manner divided into about eighteen distinct operations.

By the end of the 18th century many workshops had adopted the division-of-labor organization along with some standardization of products. Seven separate activities, for example, were involved in making Brussels lace. Even so, standardization and "mass production" of items was still minimal. Most workshops employed only a small core of laborers.

The change, when it came, was dramatic. In 1603, a British regulation stipulated that:

> no clothman shall keep above one loom in his house, neither shall any weaver that hath a ploughland keep more than one loom in his house. No person or persons shall keep any loom or looms going in any other houses besides their own.

Yet, 100 years later, English textile "factory" operators were using several large handlooms in single manufacturing rooms.

It was machinery and innovative methods, along with the availability of coal, that turned workshops and cottage industries into actual factories. Simple water-powered machines were important in the earliest factories that existed even before the Industrial Revolution. Power looms, as such, were not in use until well into the 18th century, however. France's King Louis XIV himself was the sponsor of a furniture manufactory—the "Manufacture Royale at the Gobelins"—established in 1667 and still operating today. A textile factory of 350 looms was in operation at Saint-Sever near Rouen during his reign. Louis XIV's successor to the French throne was the sole proprietor of the thriving Vincennes porcelain factory.

Gears, screws, and metal parts designed to replace wooden ones were the heart of the true factories that were transforming Western Europe by the end of the 18th century. Factory towns sprang up everywhere as technology and techniques became increasingly innovative, complex, and precise in the 19th century. As machines were almost totally substituted for hand tools, larger amounts

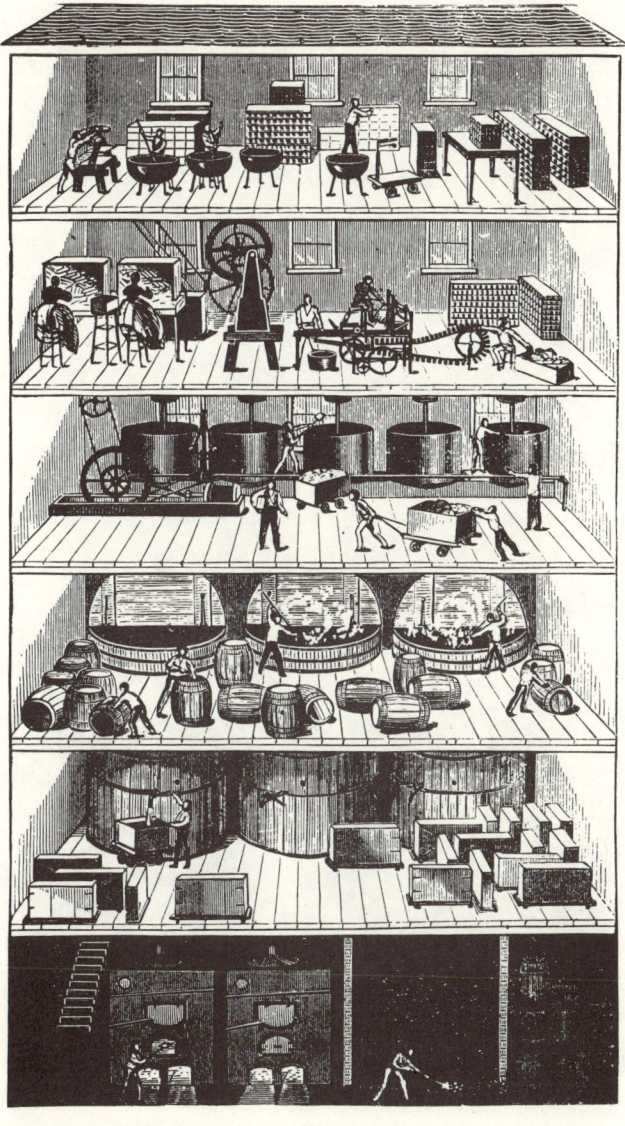

Division of labor was sometimes given a vertical structure, as in E. Morgan's Sons soap manufactory. (From Harper's New Monthly Magazine, *19th century)*

of capital were needed for organizing and operating factories. The structure of the factory system itself underwent considerable changes, and there were enormous increases and refinements in the division of labor. All factories became geared toward the cost-effectiveness of production, with which the capitalist was dominantly concerned.

Factory workers essentially became extensions of the machines they operated or the minute processes they tended to. They relinquished whatever claims to craft they might have held in the handwork manufactories or in monastic and estate workshops. While they had received little real compensation for their efforts in those early, prototype factories, they had at least received the

Modern-style factories, like this one in Lancashire, first appeared in the cloth-making industries. (After T. Allom's drawing in G. N. Wright Lancashire, its History, Legends and Manufacturers, 1842)

protection, care, and relatively humane regard of their employers. But factory workers received virtually none of these benefits from the capitalist-entrepreneurs who operated the new factories. They were given materials, tools, and devices or machines with which to work, as had their predecessors, but they remained at the mercy of their employers, as their standard of living became increasingly tied to a wage, with less and less recourse for subsistence farming to help provide for themselves and their families.

Women had always been important in the textile industries. Since those industries were the first to adopt large-scale factory organization, women became a key to the labor supply of the new factories in many other in-

Young girls like 12-year-old Addie Laird were often employed in spinning mills. (By Lewis Hine, Children's Bureau, National Archives, 102-LH-1056, North Pownal, Vermont, 1910)

dustries as well. Women also were more used to doing monotonous work in a confined space. Men continued to prefer farming and other outdoor work to indoor factory production. Along with women, young children were used to bolster the labor supply for factories before child labor laws were enacted after the mid-19th century. In England, where the arrival of the factory transformed society most suddenly and drastically, children between the ages of 8 and 12 were *short-timers* obliged to attend factory "schools." There they received a rough and very basic kind of education. By the age of 13 they were allowed to become "full-timers" and work as aides to craftworkers. The children were given the simplest of the chores. In the loom areas they served as *tenters*, stretching fabric on nails called tenterhooks, to prevent it from shrinking, for example. Women did the majority of the work that was a grade higher, while the men held the most-skilled positions.

One contemporary observer said that Birmingham, England, became a "forest of tall chimneys" during the Industrial Revolution. Many peasants, unemployed wanderers, women, and children flocked to this and

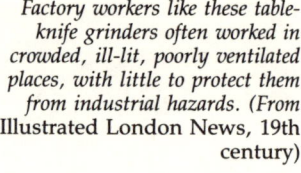

Factory workers like these table-knife grinders often worked in crowded, ill-lit, poorly ventilated places, with little to protect them from industrial hazards. (From Illustrated London News, 19th century)

18 MANUFACTURERS AND MINERS

other new urban areas in search of work. They found wages, but labored dearly for them—and under the severest and most depressing of circumstances. Poor space, lighting, ventilation, and cleanliness were chronic problems. Some workshops had to be reached by climbing dangerous stepladders. Button manufactories had low headspace and were grossly overcrowded, since only sitdown, hand-motion work was necessary. Such shops were worked by scores of young "girls creeping in under women's legs and benches" just to get to their seats. Once in place, they remained working for hours, shoulder to shoulder with co-workers, barely able to move anything other than their hands. The noisy casting and stamping shops, which needed hardly any lighting for their operations, were contained in damp and dark cellar spaces. Pits had to be dug in the dirt floors to gain headroom, and light came only from a few tiny windows or from furnace and forge fires used by workers on the night shift.

Factory workers were more than uncomfortable and inconvenienced. Many were injured or suffered serious

Factory work has often involved considerable danger and minimal protection; here workers are lowering canned tobacco into a boiler. (From Frank Leslie's Popular Monthly Magazine, 19th century)

health problems and even premature death from dangerous machines and toxic fumes, to mention only a few hazards. *Potters* frequently suffered "painter's colic" or crippling paralysis from dipping wares into poisonous lead glazes. *Match manufacturers* had severe tooth, jaw, and lung disorders from overexposure to sulphur and phosphorus. *Bootmakers* often lost an eye when stabbed by co-workers' needles—not an uncommon occurrence, given the close quarters in which they worked. Children were frequently sent to factory schools in order to get a breath of fresh air from noxious fumes, but adults received no such relief. All workers had to be fanatically punctual and keep pace with production quotas and assembly lines, especially from the late 19th century on. Any deviations or lapses were handled with loss of wages, verbal abuse, and even physical beatings in some cases (especially for children).

Despite all of the hardships factory workers endured, many did so not only because they had to feed their families, but also because they favored factory work to other alternatives. Young girls flocked from the countryside to savor the excitement of the city. They found three opportunities for employment: prostitution, domestic work, and factory work. The latter two, of course, were preferred by most, and a great many liked the independence of factory work better. In the factories, young people got to work together, socialize on breaks and after hours, and did not have to wear uniforms. Unlike *domestic servants*, they were not confined at all times to the household, nor were they the moral charges of any householders or employers. They did not worry so much about the stench and squalor of their rooms, often in or beside the factory itself, because they were free to leave when they wished. In addition, factories sponsored large annual or semi-annual social events for their workers, the winter ball and summer picnic being the most popular of these.

All things considered, though, factory work was no picnic. It was hard and grueling for men, not to mention

Workers are dwarfed by the huge Marseilles soap factory—and little protected from its hazards. (From Harper's, 19th century)

children and women. Wages were so low that laborers could rarely improve their lot, and they generally lived shortened and miserable lives. Things began to improve when England passed a series of Factory Acts beginning in the 1860's, and other countries followed its example. Essentially, legislation began to promote more safety and better health standards for all workers, and mandated shorter work hours for women and children.

By the middle of the 20th century, most industrialized nations had established some sort of minimum-wage laws ensuring that factory workers (as well as others) could not be taken unfair advantage of. Other legislation did away in stages with child labor. With the increasing complexity of technology and mechanization, it even became possible to further restrict the number of work hours for adults. The 40-hour workweek is now the norm in many countries, as opposed to 60-, 70- and 80-hour weeks at the beginning of the century. Most factories and governments now offer life and health insurance to factory workers, as well as annual vacations, sick-leave allotments, and even social and recreational opportunities on the premises.

While the lot of the factory worker in industrialized countries has improved dramatically in the last 100 years, these changes were not easily achieved, nor were they purely the result of enlightened legislation. Labor reform movements were primarily the blood-and-guts work of miners and factory workers themselves, who fought against powerful capitalists and their immense financial interests and investments, as well as their vast

The Industrial Workers of the World (IWW), here addressed by young Alexander Berkman, was but one of many labor unions fighting for better working conditions. (By Bain, Library of Congress, 1908)

political influence. For the most part, the root of labor reform in the factories has been the organization of factory workers into *unions* with the right to *collective bargaining* in their own interest.

Unions were, of course, bitterly opposed by industry management. Only bloody confrontations and paralyzing strikes slowly but steadily won the attention and support of enough *legislators* to win workers the legal right to organize and bargain collectively. Out of such efforts came the minimum wage; safety and health legislation; retirement and pension plans; and measures to stop child labor abuse and other callous management policies and tactics. The fate of factory workers has had other, even more far-reaching effects as well. Some of the major social and political movements of modern times, notably socialism and communism, have focused on the relationship between the working class and those who pay for and profit from their labor.

Today, factory workers in the industrialized countries generally enjoy comparatively high standards of work-

Before modern health and safety legislation, factory workers like this moldmaker had little protection on the job. (By Lewis W. Hine, Works Projects Administration, Dec. 1936-July 1937, 60-RP-57)

ing conditions, although workers in many industries—such as asbestos, plastics, and chemicals—still face serious and even potentially fatal health hazards every day. But, for the most part, factory workers receive relatively decent salaries and benefits programs, due largely to the successful bargaining of their union representatives. Their social status is still low, however, and their jobs are generally mundane, depressing, and unskilled. Many spend countless hours at assembly-line production, in which they repetitively perform one simple, little process—like tightening bolts or attaching standardized parts—in the overall production of an item. The psychological rewards for such labors are minimal, as a worker's efforts rarely produce anything close to a total, finished, or unique product. Lacking this satisfaction and stimulation that once belonged to the craftworker, the factory worker's only incentives are money, vacations, bonuses, and promotions.

Nor are communist countries immune from such problems. Even in nations supposedly controlled by the workers, the question of motivation for increased production continues to plague economic planners. Several communist nations have experimented with entrepreneurial incentives as a way out of this dilemma. These nations have also not solved the questions of safety and health. Workers in Western democratic countries are generally better protected than those in communist countries. There, unions—that would fight for such protections—are presumed to be unnecessary under the supposed rule of the proletariat, and are therefore outlawed. The fate of the Solidarity movement in Poland is a graphic reminder of the difficulties unions face in commmunist nations.

But as unions have gained higher wages and better benefits, they face increasing competition from nonunionized workers. Within industrial countries, businesses have often relocated to areas without strong labor movements. In the United States, for example, many factories in recent decades have moved to the South and Southwest, where workers lack a strong union tradition and will work for much lower wages and under poorer conditions.

Some businesses have even moved their plants to developing countries, where factories employ people in conditions and for wages much like those in the earliest and worst of modern industrialized times. Workers there are generally underpaid, overworked, and underprotected—and so produce goods at much lower cost than workers in unionized plants in industrialized countries.

On the other hand, workers in some industrialized countries face stiff competition from countries with more modern plants. Countries such as Great Britain and to a lesser extent the United States are operating increasingly older, less-efficient plants in many types of manufacturing. However, Japan and West Germany, whose plants were mostly destroyed during World War II, have newly built, modern plants. Japan especially has led the way in developing automated plants, where robots largely replace human workers.

As a result, the greatest concerns of many contemporary factory workers are plant shutdowns, cutbacks, and unemployment. While factory labor unions have pressed for higher and higher wages and more and more benefits, many industries have been shutting down factory operations or placing them on severe austerity budgets. Factory workers, particularly in the automobile industry, have been hard hit. Although many factory workers have been laid off in the United States, for example, many more have had to take voluntary cuts in wages and benefit programs in order to keep the factory itself open. Occasionally factory workers' groups and unions have even banded together to take over the management and operation of factories that would otherwise have closed. Because of these problems factory workers are in a very shaky situation, either looking for other types of work or being left with none at all, thus becoming burdens to the state. The many millions of employed factory workers still able to hold on to their traditional occupations, however, manage to maintain a better standard of living and safety than their forebears ever did.

For related occupations in this volume, *Manufacturers and Miners*, see the following:
    Mechanics and Repairers
    Metalsmiths
    Miners and Quarriers
    Vehicle Makers
    Weapon Makers

For related occupations in other volumes of the series, see the following:
in *Artists and Artisans*:
    Clockmakers
    Furniture Makers
    Glassblowers
    Jewelers
    Locksmiths
    Potters
in *Builders*:
    Architects and Contractors
    Carpenters
    Masons
    Shipwrights
in *Clothiers*:
    Dyers and Other Cloth Finishers
    Fiber Workers
    Shoemakers and Other Leatherworkers
    Spinners
    Weavers
in *Communicators*:
    Printers
in *Financiers and Traders*:
    Bankers and Financiers
    Merchants and Shopkeepers
    Stewards and Supervisors

in *Harvesters*:
  Farmers
in *Helpers and Aides*:
  Servants and Other Domestic Laborers
in *Leaders and Lawyers*:
  Inspectors
  Political Leaders
in *Restaurateurs and Innkeepers*:
  Bakers and Millers
  Cooks

Mechanics and Repairers

Mechanical repairs have traditionally been done by the owners and operators of the faulty tools or machines. In the ancient world, *millers* gave their *slaves* the chores of replacing broken grindstones and of repairing broken waterwheels. Sometimes the *merchants* who sold items also made repairs on them. *Cooks* usually sold and repaired kitchen utensils; *shoemakers* mended shoes; *wagonmakers* fixed broken wheels. None of these were strictly professional mechanics or repairers, however, since repairs were only a minor portion of their trade.

This situation did not begin to change until the time of the Renaissance, when Europe was flooded with many new gadgets and machines. Windmills, firearms, and the printing press all became important at that period.

The mechanics and repairers of these and other mechanisms were most typically the *engineers* who had designed them or the *carpenters* or *metalsmiths* who had constructed them. Simple repairs continued to be made by the owners and operators. But, as machines and gadgets increased in complexity, more intricate and delicate repairs required the services of an outside expert. Still, very few people made an actual business of repair work.

The colonization of the New World gave some impetus to the occupation. The North American colonies were widely dispersed, as were their major urban centers. Rural life on the vast continent was much more removed from city shops and commercial activity than it was in Europe. If householders or farmers did not know how to repair tools or metalwares or furniture, it was no simple task to take them to the *blacksmith, pewterer*, or *cabinetmaker* to be fixed. Instead, it became common practice to wait for the wandering *peddler*—who often had sold the wares to begin with—to stop by and provide this service.

While this grinder seems to have a permanent shop, many others roamed the countryside sharpening blades and fixing metal and wooden items. (By Jost Amman, from The Book of Trades, *late 16th century)*

Peddling was not a regular employment in the rest of the world, where communities typically had developed over many centuries with closer connections to urban centers. Peddlers, then, were not needed so much; in fact, strong *shopkeepers'* and *traders'* guilds effectively barred peddlers from competing with their repair activities. American merchants and craftspeople also tried to inhibit the growth of peddling, but succeeded only in keeping it away from the major cities. The Yankee peddler roaming the remote countryside became a prototype of the *mechanic* and *repairer*. The *tinker* mended metalwares, especially pewter, while other peddlers earned special reputations for fixing shoes, grinding blades, or repairing furniture. Most good peddlers could do a little bit of everything, and it soon came to be expected that peddlers should be able to fix what they sold.

The real specialization and modern foundation of the mechanic and repairer occupations grew out of the Industrial Revolution and the great machine age. Technology and the mechanization of production became so com-

Traveling knife grinders and tinkers, with small portable furnaces for metalworking, carried out repairs for country folk for many generations. (By W.H. Pyne, from Picturesque Views of Rural Occupations in Early Nineteenth-Century England*)*

plex and sophisticated that the people who owned and operated machines did not necessarily know how to maintain or repair them. The invention of the steam engine, electric motor, and vacuum compression began to revolutionize machines as the 19th century progressed.

The 20th century has brought an unprecedented mass production and distribution of complicated machines, gadgets, computers, and even toys. Today, relatively few owners or operators of machines know how to maintain or repair them. In many cases they have to be carefully guided or specially trained even to install or use them. For example, many people need "driver's education" to learn to operate the motor vehicle they have purchased, or close consultation with a representative from the company that has sold and installed a computer system in their business office. Even simpler machines like clothes washers and dryers usually have to be installed by a professional mechanic.

Mechanics and repairers today work in virtually every major type of manufacturing company. They are retained

In modern times, the automobile mechanic's shop has become a familiar and welcome—if not beautiful—sight along the road. (By Arthur Rothstein, from The Depression Years, Dover, 1978)

to fix returned or damaged items and to do routine maintenance. They sometimes even assist in the mechanical design of products. Some work for large retail stores, which offer to service the goods they sell. Others work independently in their own shops or travel to people's houses, like the peddler and tinker of days gone by. With so many machines to be worked on, though, most mechanics and repairers have done away with "house calls," except for machines such as refrigerators or clothes washers, which cannot practically be brought to a shop. Many *television repairers* will still visit homes, but the tendency has increasingly been to bring most work into the shop. Most *automobile mechanics* work only in their own repair stations, though they will pick up broken-down vehicles with tow trucks, and a few even make a business of doing the work at the customer's home.

Mechanics and repairers have traditionally been men, but women have shown increasing interest in this type of work in recent years. A wide variety of fields are available to these workers, some involving little training and others involving extensive training and formal education in mathematics and science. Military and aerospace mechanics have been very important since the mechanization of combat. Transportation is another major area, with trains, airplanes, jets, trucks, and automobiles all demanding specialized treatment. The computer and electronics industry also makes extensive use of professional repairers, although workers in this field (as in others that demand especially high degrees of training and knowledge) are often referred to as *technicians*, rather than mechanics or repairers. Small machine and appliance mechanics include those who repair and maintain typewriters, fans, vacuum cleaners, toasters, bicycles, firearms, power tools, and welding equipment. Some of the larger machines and structures worked on are aircraft, pipelines, motor boats, farm and construction equipment, and printing presses. (Many printing and publishing companies employ a regular *press maintainer*.)

Some of the occupations in the field also involve work that is experimental or research-oriented. These include specialists in rocket, missile, and aerospace mechanics. In this category are such professionals as *flowmeter testers* and *certification mechanics*, who inspect and repair rocket-testing equipment and control panels. They frequently devise and fabricate replacement parts and may help plan and construct new equipment.

The mechanization and industrialization of society, added to the technological revolution that swept the world after World War II, have led to a vast expansion of the profession. The mechanics and repairers of today provide truly essential services to a society that is dependent on the smooth operation of high-level technology and mechanization.

For related occupations in this volume, *Manufacturers and Miners*, see the following:
    Metalsmiths
    Vehicle Makers
    Weapon Makers

For related occupations in other volumes of the series see the following:
in *Builders*:
    Carpenters
in *Clothiers*:
    Shoemakers and Other Leatherworkers
in *Communicators*:
    Printers
in *Financiers and Traders*:
    Merchants and Shopkeepers
in *Restaurateurs and Innkeepers*:
    Bakers and Millers
in *Scientists and Technologists*:
    Computer Scientists
    Engineers

Metalsmiths

Metal has been responsible for much of the technological progress that humans have made from prehistoric times to the modern industrial world. It changed the entire process of war through the development of swords and firearms; the methods of agriculture with plows and hoes; haulage and travel with horseshoes. It was central to the success of printing, rail and automotive transportation, and mass production factories. And it has made possible the large-scale urbanization of society. So central has metal been to human progress that *archaeologists* have even used different kinds of metals to mark the stages in the development of civilization—from the Stone Age to the Copper Age, to the Bronze Age, to the Iron Age, and modern times.

For a material with such astounding social impact, its

producers and shapers have not always received their due respect. On the contrary, many have poured their sweat and blood into a life of long hours over almost unbearably hot furnaces and heavy, wearisome manual labor such as hammering and casting for little material or psychic reward. Yet many others, who had the skills and fortunes to venture into the large-scale production of metal, have prospered. Some have been numbered among the richest and most successful entrepreneurs in history.

Metalworking is an ancient profession that existed on a limited scale during the many centuries before the Industrial Revolution. No one knows who the first metalsmiths were, but most of the peoples of antiquity used metalwares. Copper was separated from ore in Iran before 4000 B.C., bronze casting was popularized by the Egyptians after 3000 B.C., and iron production helped shape the powerful Hittite empire. Gold and silver were worked by many ancient cultures, from the Japanese and Chinese to the Native American. Lead and tin have also been worked since these early times. Later, even steel was

This is a modern depiction of an iron-smelting furnace such as might have been used by prehistoric peoples. (From Peoples of the World, *19th century)*

worked by the Moslems for the manufacture of their famous damascened, or Damascus, swords.

Many of the early *metalsmiths* were *slaves* and *soldiers* who accompanied royal troops to the mines where they helped to extract crude ores or simply to pick them up—many metals simply lay on top of the ground, especially in desert hills. They then constructed earthen furnaces at the site, where metals could be extracted by heating (*smelting*) and then shaped by pounding, stretching, or pressing (*forging*). These slave and soldier smiths typically worked without pay and did so in service of the priesthood, court nobility, or king. Most of this metalworking was done for the purposes of accumulating precious metals, adorning palaces and temples, and dressing up elite military officers and courtiers. Royalty used silver and gold for the eating and drinking utensils in their households. Some very prosperous town craftspeople could even afford metal tools.

Iron was the only metal that could be put to rugged, practical use, and its effect on history has been greater than that of gold, silver, copper, and bronze all put together. The first iron *smelters* probably obtained crude or wrought iron as a by-product of gold processing. The Hittites were the first to work iron in appreciable amounts, realizing the significance of this hard material for hunting and fighting weapons. Between the 15th and 12th centuries B.C., *ironsmiths* began to construct pottery *blooms* (also referred to as *bloomeries*)—furnaces heated by wood or charcoal, designed for the low-temperature reduction (melting to remove impurities) of soft wrought iron. Because of insufficient blasting or blowing techniques to fan the flames in the furnace, early ironworkers could only reduce the ore to a pasty mass, but could not heat it sufficiently to melt it. Iron could not, therefore, be cast like bronze until better techniques were devised.

The Egyptians and Hittites both sponsored massive mining and metalworking industries. While slaves and soldiers provided the bulk of the labor, skilled metalsmiths were beginning to be recognized for their special talents. Some were commissioned as foremen and project designers, increasingly so as the complex nature of these metallurgical processes was recognized. As smiths performed many seemingly miraculous tasks, such as separating silver from gold or lead from silver, they were frequently regarded with great reverence for their presumed mystical insights. The early ironmasters were those brave enough to work on ores extracted (at that time) solely from fallen meteorites—what the Egyptians fearfully referred to as "stones from heaven." These masters of the black metal—these so-called *blacksmiths*—were a highly honored lot in Hittite culture, forming elite, close-knit brotherhoods.

By the second century B.C. the Chinese and Indians had developed a process for converting iron into soft steel, and some had even stumbled upon methods for making hard steel. Known as *seric iron* in China, and *wootz steel* in India, hard steel was used for the Damascus swords and other highly prized superior bladed weapons. Its makers carefully guarded their secrets for processing this steel, and handed it down only within the family. Special cults sometimes grew up around such metallurgies, and alchemy itself arose from the desire to understand these secrets.

As metalworking and smithing became more prosperous, it was taken up by families and sometimes partnerships or small companies. The mines of ancient Timna, in Israel, were attended by men and boys who squatted on the ground for long hours, crushing copper ore with granite rocks. At Abu Matar, a little farther north, underground casting rooms were built in caves for the work of the *founders*—the people who melted metal and poured it into a mold to create the desired object. Founders typically consisted of a cohesive group of families working together. *Forgers*, often called blacksmiths, worked at smallish shop furnaces called *forges*. There they heated metal and then generally beat or hammered it into the desired shape.

Not all metalworking was done on a small scale, though. There was a major copper industrial site at Tal-I-Iblis in Iran as early as 4000 B.C. There *forgers, founders*, and other smiths and laborers lived in tiny one-room dwellings and worked the ores in such great quantity that there was enough left over for overseas exports. The Egyptians operated 24-hour furnaces, so that the fire did not have to be rebuilt every day, which would have wasted fuel, labor, and production time.

Other metalworkers were wanderers, welding and mending tinwares and copperwares, as some Gypsy *tinkers* still do today, especially in Eastern Europe. They also did forging and founding, carrying their own metal bars or sheets with them to do on-the-spot, custom work for their wealthy clients. They made their own smelting furnaces and bloomeries by gouging ruts in the sides of hills, where wind pockets might swirl to keep the charcoals burning hot. A copper mine in the Koszeg district of Hungary had a large metalworking shop where finished goods were forged, or where melted copper was cast into ingots (rough masses), shaped into bars, and sold to the wandering smiths. Many of these roaming

The Chinese were skilled at refining metals from very early times. (From The Exploitation of the Works of Nature, *1637 A.D.)*

METALSMITHS 39

metalworkers became masters in making tools and weaponry, although they continued to serve an elite clientele in this capacity, as they did in the crafting of jewelry and decorations.

Metalworking became important early in history and in virtually every corner of the earth. The Chinese and Japanese worked wonders with bronze, as did the Native South Americans with gold. The fifth-century B.C. Greek historian Herodotus claimed that Ethiopian kings had for a long time bound prison gangs with gold chains. After the Hittites had popularized the use of iron, metals became the focus of society, no longer just for the wealthy and elite nobles, courtiers, and warriors. Early toolmakers created a virtual revolution with the development and dispersal of the axe and the plow. The nomadic, restless life-styles of many peoples would eventually be changed by such simple tools and farm implements, which permitted the clearing of forests and helped to break and till the soil. Culture, by and large, became more sedentary because of the labors of the metalsmiths, and the great civilizations of antiquity soon depended on them.

Woodworkers and *carpenters* could hardly ply their trades without chisels and saws, which were used in the manufacture of so many fine homes, temples, and public buildings, as well as the furniture that went into them. Iron wheels made cart and wagon transportation practical for the first time—thus aiding the farmer almost as much as the plow did and stimulating trade and commerce between peoples. Waterwheels and the secondary gearing systems that were often attached to them could not have been possible without metal. And without the waterwheel, lumber and gristmilling, mass production, and heavy manufacturing could never have evolved.

Not all cultures in the world developed metal technology so early. In the years before 1000 B.C., picks, rakes, and shovels were still being made out of antlers and the shoulder blades of oxen in the British Isles. Tribal peoples in the interior of Africa, as well as in North America, were still making their tools out of blunt and highly unreliable stone and flint. But the great civilizations of China, Greece, and Rome were based squarely on the working and use of metal tools and implements.

From the days when the Egyptians operated huge copper factories to the extensive founding and faggoting (hammering and welding of metal bars) of arms and armor during the Imperial Age of the Romans, metalworking as an occupation really underwent very few changes. Tongs and foot-operated bellows had been used as early as Egyptian times, although the Romans appear to have been the first to use water-powered bellows to keep the bloomeries burning hot. Blowpipes were also used to fan coals. The Egyptians had used long clay-covered pipes for this purpose, and later smiths used rawhide blowers. Metalsmiths would become easily exhausted working these blowers, and this chore was frequently turned over to slaves or boys working in relay teams so that the blowing never stopped.

Smiths in Eastern Europe were expert at faggoting iron tools—that is, hammering and stretching them into shape by alternately heating and hammering the wrought iron. Such work required always being close to the bloomery, though, and many blacksmiths suffered recurrent heat exhaustion and dehydration. Founders, too—although they worked with softer metals such as copper and bronze that could be cast once melted—spent long hours at the bloomery testing the melted masses and constantly skimming impurities that rose to the surface during the heating. Most metalworkers ended up having skin like "crocodile hide" and worked up such a constant sweat through every pore of their bodies that they could not help but stink like "the roe of fish," as an old Egyptian text puts it. They often developed cramped and arthritic legs and backs from laboring in heated premises that had damp clay and sand floors to absorb the fiery spills of ash and metal and prevent fire hazards.

As some of the first practical *chemists*, metalsmiths were also avid experimenters, always attempting to create new alloys (combinations of metals) or to establish better methods of assaying (separating metals from ores). As practical experimenters with little real scientific knowledge, they sometimes suffered from their trial-and-error dabbling, which could produce very toxic chemical reactions, base (nonprecious) materials, or gases. Many metalworkers were killed by working with copper and arsenic, and countless others must have suffered chronic side effects or degenerative disorders from such contact.

Few metalsmiths were educated or even very thoroughly trained, except by their families or employers. Still they could claim respect because of the mystery and technical nature of their occupation. As chemist, *goldsmith*, and *jeweler*, the highly accomplished metalworker was closely associated with and often protected and patronized by the priesthood, nobility, and royalty. Traveling metalworkers had less prestige, although their visits were eagerly awaited by artisans and farmers, who desperately needed their tools, and by warriors and nobles, who desired adornment, weapons, and wheelwork. These wandering metalworkers were, of course, virtual "one-man shows," but there was little division of labor even in the larger towns or mine workshops. A metalworker usually knew the basics of bloomery maintenance, founding, forging, welding, soldering, and riveting, but seldom specialized in any one aspect. Dwarfs were frequently used by the Egyptians to do fine and finishing work because it was believed they had more manual dexterity than the average worker; slaves were used as *fillers* of the blooms, *blowers* of the flames, and general *porters* of raw materials.

Metal continued to be mined, smithed, and smelted in relatively small amounts during the Middle Ages. In the ninth century A.D. metalworking was done largely by peasants as part of their feudal obligation to their landed lord, who had had the ores mined. Some enterprising citizens even mined and smelted their own ores, but they had to pay a substantial portion of any profits from this activity to the lord on whose land the mine was worked. The monasteries were the most important metalworking centers for several centuries. Both the Cistercians and Carthusians employed lay brethren known as *conversi* to supply much of the labor. The Abbey of Clairvaux in Burgundy alone had a dozen ironworks, and in the late Middle Ages was the leading single iron producer in the world.

Although peasants had done much of the labor involved in metalworking up to this point, by the time of the Renaissance and the later Commercial Revolution, skilled metalworkers came into far greater demand. The increase in commercial activity and trading brought a greater need for metals. But as the feudal structure of society broke down, peasants were no longer so readily available for metalworking. Paid metalsmiths filled this void to some extent, but mechanization did so even more. The industry became increasingly reliant on machines to

This rare woman metalsmith is making nails for a cross. (From Holkham Bible Picture Book, *c. 1330)*

do the heavy and burdensome labor and was more dependent on the skilled smelter, forger, and founder to work the increased volume of metal being produced.

General metalsmiths had tended to work in virtually every aspect of the trade, but now some began to specialize. Once only general *farriers* had worked iron. But by the time of the Renaissance the *blacksmith* had become primarily the *horseshoer*, as a special offshoot of the farrier's occupation. Metalworkers, however, tend to specialize in a particular metal. As the Commercial Revolution began in about the 16th century, fairly strong metalworking guilds had already existed for some time: *bronzeworkers, silversmiths, goldsmiths, tinners*, and *blacksmiths*. They formed cohesive brotherhoods of professionals, who handed down carefully safeguarded trade secrets exclusively to guild brethren, and sometimes only to family members. Yet, within these specializations, derived according to the material each group worked with, smiths were still essentially complete master metalworkers. They frequently did their own smelting, forging, founding, and finishing. Each one was a little bit of a scientist, an alchemist, a philosopher, a manufacturer, a craftsperson, a tinker, and a merchant.

Metal production was beginning to become more mechanized about this time to better serve a rising demand. As early as the 14th century, bloomeries in the Rhineland had become larger. This development was made possible only by increasing the power and consistency of the blasts through the use of water-powered bellows, common features of the industry by the time of the Reformation. This increased heat intensity allowed blacksmiths and farriers to run iron into hollows called *sows*, thereby producing the first form of *pig iron*. Blast furnaces eventually displaced bloomeries, thereby maximizing the available energy and process efficiency. By the end of the 16th century this innovation had led to the pouring of iron by the ton instead of the hundredweight. Yet demand for iron continued to rise. The wars of the 16th century, the need for increased agricultural output,

Blacksmiths were vital artisans in both farm and city, for they shoed (and often treated) horses and repaired wagon wheels. (By Jost Amman, from The Book of Trades, *late 16th century)*

the rise in town crafts and piecemeal manufacturing, and the increased amount of inland travel and commerce all required arms and armor, plowshares, hand tools, and horseshoes in unprecedented quantities.

The metalworking professions had become a part of a very large and prosperous industry. However, farriers were employed only on a seasonal basis. And, indeed, the industry itself remained somewhat seasonal right up to the Industrial Revolution. Bad roads made it difficult to transport ores from the mines in the winter, and waters that had iced over in cold weather or dried up during droughts could hardly turn the waterwheels that operated bellows and mechanical hammers. Still, there was a greatly increased activity in metalworking, and part-time farriers began to give way to more and better-trained professionals.

A division of labor—in which metal was partly worked by one person and finished by one person or more—began to appear at this time. Heavy metalworking was done increasingly at *forgiae errantes*, workshops located near the mines, where metal was smelted and cast into ingots. These were shipped to the traveling and town smiths and founders. With the rise of capitalism, large-scale financial funding made it possible to combine many of these smaller forges into massive factory sites, often located in the cities rather than the country to make better use of labor and to have more direct contact with customers. With fewer but larger forges, energy was greatly conserved, thus somewhat alleviating the major problem with power shortages in this industry. It also led to some degree of specialization among smelters.

Many metalworkers were still wanderers between the 16th and 18th centuries, but many more were employed in large forges or in city shops. The shops consisted of rows of stalls in which *blacksmiths, pewterers, tinsmiths*, and others noisily plied their trades by smelting and refining hardwares in full public view, to the annoyance of many. Metal from ingots or scraps was hammered manually and sometimes mechanically; sheet and wire were drawn; and tools, weapons, and farm implements were cast, faggoted, welded, soldered, and riveted in such shops. In his 1603 *Survey of London*, John Stow took note of the streetside din created by the activities of the metalworkers of his time:

> [the street] is possessed for the most part by founders, that cast candlesticks, chafing-dishes, spice mortars, and such like copper or laton works, and do afterwards turn them with the foot, and not with the wheel, to make them

The work of these scale makers had to be every bit as delicate and fine as that of jewelers, clockmakers, and locksmiths. (From Diderot's Encyclopedia, *late 18th century)*

smooth and bright with turning and scrating (as some do term it), making a loathsome noise to the by-passers that have not been used to the like...

Many a smith worked both in the shop and on the road, sometimes leaving others behind to tend to shop business. As capitalists became involved in the industry, larger, more mechanized, and more efficient factories and workshops appeared. One great banking family, the Fuggers, owned copper and silver furnaces in Hungary and sent ingots or bars, rods, and pipes all the way to Italy and the Baltic regions.

However much the status of the metalworking industry may have changed from the Renaissance to the Industrial Revolution, the workers themselves were doing essentially the same work under the same conditions that had prevailed since antiquity. They were now aided by hearths and blast furnaces, along with water-powered bellows and hammers, but they still did mostly exhausting, dirty, and hot manual work. New metals being worked—including zinc, cobalt, and nickel—created new opportunities, but also new dangers, since there was little real knowledge of the materials being worked so zealously. Young girls in mining districts freely applied deadly arsenic dust to treat facial acne; and mercury was used in folk medicines for a long time as a cure for syphilis. Metalworkers themselves were subjected to countless hazards and toxins on a regular basis.

A famous treatise dealing largely with the subject of the metal industry of the early 16th century was Vannoccio Biringuiccio's *Pirotechnia*. In it he described the general lot of the metalworker:

> He who wishes to practice this art must not be of a weak nature, but must be strong, young and vigorous. . . . Nor do I doubt that whoever considers this art well will fail to recognize a certain brutishness in it, for the founder is always like a chimney sweep, covered with charcoal and distasteful sooty smoke. . . . To this is added the fact that for this work a violent and continuous straining of all a

man's strength is required, which brings great harm to his body and holds many definite dangers to his life. In addition, this art holds the mind of the artificer in suspense and fear regarding its outcome and keeps his spirit disturbed and almost continuously anxious .... But, with all this, it is a profitable and skillful art and in large part delightful.

As the Industrial Revolution got under way, metalsmiths were just solving their energy problems. Charcoal, used extensively in the northern countries of England, Sweden, and Russia, had become excessively expensive. Its production was being increasingly restricted by legislation designed to protect shipbuilding industries, which were necessary for naval power, but competed with the *charcoalmakers'* dwindling lumber supplies. Moreover, deforestation was posing serious environmental problems, and the age of massive and wasteful charcoal burning in the metalworks industry was near its close. Coal ultimately came to the rescue, supplying abundant and cheap fuel to metalworkers in Great Britain, where the Industrial Revolution had its greatest original impact; and to those in America,

As demand for iron began to grow, especially in the building trades, forges became ever larger. (From Diderot's Encyclopedia, *late 18th century)*

Sweden, Russia, Germany, and France, where the age of mechanized production was also strongly felt.

The new age of mechanization and industry gave the profession of metalworking its greatest thrust in history. Better processes and innovations in mechanics, along with the advent of coal, made metalwork more widely available. In England the base price of pig iron was actually cut in half between 1728 and 1802. Heavy metals—especially cast iron—soon came into great and unprecedented demand by makers of rails, bridges, and even engine cylinders. The working parts of engines, though, required the greater strength, hardness, and resiliency of steel. For centuries steelmaking had been kept a carefully guarded secret by those few, exclusive groups who knew it. René-Antoine Réaumur had discovered in the early 18th century that steel was essentially iron ore reduced until it had just the right carbon content. Yet it was not until enough technological improvements were made that *steelmakers* could claim to be a separate profession from the farriers. Compressed and hot blasting made wrought iron (a grade above cast iron) much more available. But it was the Bessemer converter and its later refinements that finally allowed the cheap, mass production of steel by the latter part of the 19th century. For quite some time, *ironmasters* fought these developments in steelmaking—and for good reason. Once cheap steel was developed, it quickly became the king of building, construction, and precision metals. **Steelworkers became the most important of all metalworkers.**

Iron was the basic industrial metal of the 18th and 19th centuries. Even in Colonial America, Andrew Oliver reported that there were 41 forges and 14 furnaces in Massachusetts by 1758. Thirteen years later, a 600-acre ironworks was constructed in Salisbury, Connecticut. A variety of items was produced there by forgers, founders, and blacksmiths; included forge hammers, blacksmiths' anvils, chimney backs, and gristmill hardware. The Trenton, New Jersey, ironworks established by Isaac

Harrow in 1723, added a planing and plating mill 11 years later and produced hand tools, shovels, axes, and frying pans. By 1772, renamed Trenton Blistered Steel, it provided *ironmongers* with the raw material of their trade "either in the blister or neatly faggoted." In addition, it also sold small rods and bars to town merchants, who had them worked by local blacksmiths into carriage springs, saws, scythes, and other wares.

The 19th century saw a much wider application of iron and then steel, first in England and then in the United States. Railroads had revolutionized transportation by the middle of the century, and large joint-stock companies and eventually corporations got into the business. With the influx of capital, the metalworking industry became marked by ever-increasing technological innovations and new machinery and by greater sophistication in chemical processing. As a result, iron and steel were to become the prototypes of mass production in the metal industry. It was in the iron and steel forges and foundries that the modern division of labor in metalworking was born. The sudden change in the tempo and structure of the profession is marked in a Shropshire, England, ironworks, as described in an article on "The Black Country," printed in *The Edinburgh Review* in April, 1863:

> The smelting furnaces are the centre of activity, and to them tramways and railways converge, bearing strings of trucks loaded with materials; and the "bridge-house"—as it has been called, because it connects the top of the furnace with the furnace yard—is full of men breaking the limestone which serves for flux, and wheeling the calcined [heated] limestone to the "filling holes."
> Under the furnace-manager the charge of the upper part of the furnaces belongs to a contractor called the Bridge-Stocker. He employs a gang of men, women, and boys, and also keeps horses, for the purpose of supplying the furnaces with the necessary materials; and as much depends on his care and regularity, it is found best to give him an interest in the work by paying him so much

per ton on the produce. The office of the "fillers" who work under him, requires watchfulness. They relieve each other by turns; night and day, with unremitting regularity, the furnaces must be fed. The work is hard, but ought to be unattended with danger. The "filling holes" or orifices by which the materials are poured down the throat of the furnace are not larger than is necessary for the purpose; a man who was "in liquor" would not be suffered to remain at the post, but man is ever making danger for himself where none exists.

Besides the *fillers*, the *bridge-stocker*, and the *furnace-managers*, there were the *keepers* and the *stocktakers* who:

prepare the sand, from the moulds, super-intend the casting, weigh the pigs [iron], and remove "the cinder." At casting-time their situation seems full of peril, but they rarely receive any injury, though they may be seen skipping about among rivulets of molten metal with more indifference than a tidy housemaid shows to the water with which she is washing the door-step; and they flit about among sparks and burning fragments of fuel as unconcernedly as a harlequin jumps through a blaze of squibs. It might be supposed that their eyes must be affected by the heat and the glare of the iron fluid; but we cannot find, on inquiry, that they are subject to blindness, or even to premature decay of sight.

The British *steelworkers* were the earliest and the best for some time. American corporations imported steelworkers from Sheffield, England, in the 1870's, but in just a few short years the United States was by far the world's greatest steel producer. The skill of these steelworkers was considerable, and their importance cannot be underestimated. Andrew Carnegie, who became one of the wealthiest and most-successful entrepreneurs in history through his exploitation of the steelmaking industry, aptly noted that "the nation that makes the cheapest steel has the other nations at its feet."

Even after the advent of industrialization, some metalworkers like these coppersmiths (coppersmiths) continued to work on a small scale. (By W.H. Pyne, from Picturesque Views of Rural Occupations in Early Nineteenth-Century England)

The process for converting molten pig iron into steel by the removal of impurities required great amounts of heat and therefore energy. Coal had become a cumbersome and expensive way of supplying this energy. After the discovery of oil and its applications in the second half of the 19th century, much of this energy came from oil.

Not all metalworking in the 19th century was done in mass-production factories, nor were all metalworkers suddenly taken into the great division-of-labor organization. The church-bell foundry of Messr. Mears in Whitechapel, London, was operated with a staff of workers who jointly converted tin blocks from the mines and copper scraps, mostly from old ship-sheathing, into finished bells. These *bell-founders* melted the metals together in a reverberatory (a furnace in which flames are applied directly to the metal, and not just beneath it), and then poured it into a huge cast. The cast, too, was made at the foundry in a great sand pit, filled in the center with brickwork to form the hollow of the bell, and worked precisely to the desired shape with clay molding. Even the sound tone was carefully adjusted by these skillful masters, and some bells were made in sets having precisely regulated intervals of tones and semi-tones. These adjustments were made by reducing the thickness of the bell where the hammer struck. This incessant chipping away of the metal with sharp-pointed iron hammers made the bell-foundry the noisiest establishment in all of London—with the sole exception, it is said, of the steam-boiler factory.

Coppersmiths and *braziers* (brassworkers) were still craftspeople when steelworkers and ironworkers were becoming factory workers. They often remained associated in guilds. Although guilds were outmoded and far too restrictive for most types of 19th-century manufacturing, they served the brass workers and copper workers by keeping their traditional secrets mystical and revered. In Germany, bell-casting was often still an occasion for great ceremony, as it had been for centuries. The bell-founder entertained a select circle of friends, associates, and community leaders, after allowing them to witness the casting itself. Prayers were shared in thanks for the craftsperson's excellent skill. Of course, the main reason that nonferrous (non-iron) metalworking remained a skill, while the iron and steel industries were rapidly being organized into huge factory systems, was that there was far greater demand for iron production. A handbook published in 1811 indicates that at least founding and finishing within copper and brass manufacturing were beginning to be differentiated as separate occupations:

> Some of the articles manufactured by the working Brazier are beat out with the hammer, and united in their several parts by solder; others are cast; those which are cast belong to the business of the Founder, except the polishing and finishing, which require the art of the Brazier. The working Brazier has need of strength, and if he would excel in his profession, he should possess ingenuity, to finish his work with taste.
>
> The Founder is employed in casting a thousand different articles in brass; for which purpose he has models of the work designed: to these he fits the mould in which he casts his metal. He rarely designs anything himself, and his chief skill lies in melting the brass, and running it into the mould evenly. The Founder requires a strong constitution to undergo the heat of immense furnaces; he may earn thirty shillings per week; but it frequently happens that he spends a large portion of it in porter [a beer].
>
> The Coppersmith makes coppers, boilers, and all manner of large vessels for brewers, distillers and others. His work is very laborious, and the business is the most noisy of mechanical employments. The wages of the journeyman are equal to the powers of body required in the operations.

By the middle of the 19th century, there was a fair blend of workshop and industrial production. Sometimes craftspeople were commissioned by the foundries to produce finished products of raw material in return for a payment plus more working material:

> The system of the manufacture of hardware in Birmingham is peculiar, and presents a striking contrast to that adopted in Manchester and other large manufacturing places—the operatives are themselves the manufacturers. Hiring a workshop in which steam-power is laid on, and which is specially fitted up by the

owner of the building, in which many such workshops are contained, the artisan plies his peculiar trade, manufactures his articles, carries them home to the merchant, and receives the weekly payment for them, which enables him to procure fresh materials, and proceed in the ensuing week with his regular labours. A very large proportion of hardwares is thus manufactured, but this system is not universal; and regularly-organized factories, employing a large number of workpeople, and possessing all the distinguishing features of a great producing establishment, exist, and are in active operation.

While metalworking was turning into a large, organized industry, many a workshop was still operated in the time-honored tradition of handcrafting and total production of wares, particularly in rural areas. Although the iron and steel industries were the first to be incorporated into mass-production systems, even traditional forms of iron-working persisted. As late as 1841 Longfellow was writing of "The Village Blacksmith":

Under a spreading chestnut tree
The village smithy stands;
The smith, a mighty man is he,
With large and sinewy hands;
And the muscles of his brawny arms
Are strong as iron bands. . . .

Week in, week out, from morn till night,
You can hear his bellows blow;
You can hear him swing his heavy sledge
With measured beat and slow,
Like a sexton ringing the village bell
When the evening sun is low. . . .

In every hamlet and town, local blacksmiths had their shops, where they forged new instruments, tools, and especially horseshoes, in the days when horses were central to both transportation and farming. With the arrival of motor-powered farm machinery and automobiles in the late 19th and early 20th centuries, however, blacksmiths gradually decreased in both numbers and importance. Some made the transition to the motor age, transforming their blacksmith shops or liveries into automobile repair shops and gas stations. Others moved elsewhere in the metalworking industries. But a few continued to work in the old way, especially in rural areas, where horses were still important for farming, riding, or hunting. In the late 20th century, some of these have combined modern technology with the old wandering role of the tinker; these mobile blacksmiths carry their forges and bellows in the backs of small, open trucks, bringing their services by appointment directly to the people who desire them.

The 20th century has seen metalworking become one of the largest and most important industries in the world. There are still many craftspeople—blacksmiths, coppersmiths, tinsmiths, and the like. But with the heavy demand for artillery in the world wars; the development of air, space, and automotive travel; and the rise of huge cities lined with skyscrapers and surrounded with

In modern times, metalworking became a heavy industry, in which highly skilled workers did difficult and dangerous jobs. (By A.R. Waud, from Harper's Weekly, *February 3, 1866)*

bridges and rampways, most contemporary metalworkers have had to settle for employment within huge manufacturing centers, where the division of labor is precise and detailed, and where craft as it relates to the total fabrication of a whole piece is a bygone concept. Nonetheless, many metalworkers today have highly skilled jobs, in which considerable amounts of training and expertise are required to master the specific phases of production in which they are situated. They often receive good pay for their efforts, and their labors produce metal products vital to the modern industrialized world. Still, as blue-collar laborers, most enjoy a social status no higher than that of other general laborers and factory workers. While metalworking has traditionally been a male occupation, women are now beginning to enter the field in small but noticeable numbers.

Metalworkers are not everywhere equally in demand, however. This has been particularly noticeable in the second half of the 20th century. The once-dominant American steel industry, and other older industries in the Western world, failed to modernize their aging factories in the decades after World War II. By contrast, Japan and West Germany, whose industries were mostly destroyed during that war, rebuilt with new, up-to-date factories, which produce steel far more economically. With their countries taking an increasing share of the steel market, Japanese and West German steelworkers have been far more fortunate than their American and English counterparts, thousands of whom have been laid off as their companies have cut back production or closed down altogether.

The greatest amount of mass-production metalworking done today is in the iron and steel industries, although aluminum processing has also become important in recent years. Copper, brass, tin, silver, gold, and other metals are also mass-produced. But they are less used for machinery and construction and so represent much smaller manufacturing businesses. Most of the handcrafting metalworkers today deal with these nonferrous metals, although blacksmiths are still finding a small amount of work.

There is an almost-endless list of occupations dealing with the mass production of metal—mostly iron and steel—products. These include the following: *Forgers* in industry today shape metal by heating, hammering, and pressing. They, in turn, include *drop-hammer operators*, who manipulate various types of hammers, which are used to forge metal parts according to work-order specifications; *forging press operators*, who use closed-die power presses to mold metal forgings; *hammersmiths* and other heavy forgers, who shape heated metal stock into forgings, employing power hammers or open-die power presses, and using knowledge in both metal-working techniques and the physical properties of metals; and *extruder operators*, who operate extrusion presses to shape hot metal into bars, tubes, rods, and structural shapes.

Founders manufacture metal castings in molds or other receptacles. They include *furnace operators*, who work with and maintain the furnaces used to heat metals to the melting point; die-casting machine operators, who set up and operate die-casting machines used to create parts to specification; *molders*, who form the sand molds for the production of metal castings, using hand and power tools, patterns or match plates, and flasks, and employing a knowledge of metal and molding sand properties and characteristics; *machine molders*, who set up and operate molding machines to form sand molds; *casting carriers*, who carry or push (on monorails) ladles of molten metal, which they pour into molds; and *coremakers*, who operate machines that make sand cores used in metal casting.

Other major occupations in the metal-working industry include the *lay-out inspectors*, who inspect castings, forgings, or cast patterns for conformance to blueprints; *mill rollers*, who set up and operate structural mills to roll slabs, plates, and structural shapes;

roughers, who operate roughing mills to reduce steel billets, blooms, and slabs to specification, using knowledge of rolling techniques and the physical properties of steel; *finishers*, who set up and operate roll stands to roll steel objects such as bars to specified thickness, shape, and finish; *press setters*, who use mechanical and hydraulic presses to produce metal parts, such as bearings, filters, and gears, according to specifications; and *scrap balers*, who operate machines that compress and bind (usually into balls) scrap metal for salvage and reuse.

Many workers are employed in various and general laboring capacities as *conveyors, pilers, loaders, markers,* and *rackers. Machinists*, both automotive and experimental, set up and operate machine tools, and fit, repair, assemble, or design metal parts, mechanisms, and machines. *Patternmakers* form the metal foundry patterns and prototypes through various manufacturing processes and through blueprint designing. *Riveters* operate machines and tools that join metal parts into flexible joints or solid pieces with rivets. *Welders* join together nonmoving fabricated metal parts. Some do this through applying heat from the resistance of the base metal to electric current (resistance welding). Others use arc, gas, electron, or laser beams, or other means to the same end. *Brazers* (different from *braziers*, meaning brass workers) and *solderers* weld metals by heating the base metals and supplying a filler metal to complete the joint.

For related occupations in this volume, *Manufacturers and Miners*, see the following:
    Factory Workers
    Mechanics and Repairers
    Miners and Quarriers
    Power and Fuel Merchants
    Vehicle Makers
    Weapon Makers

For related occupations in other volumes of the series, see the following:

in *Artists and Artisans*:
    Jewelers

in *Builders*:
    Architects and Contractors
    Carpenters
    Construction Laborers

in *Communicators*:
    Printers

in *Financiers and Traders*:
    Merchants and Shopkeepers

in *Harvesters*:
    Farmers
    Hunters

in *Leaders and Lawyers*:
    Political Leaders

in *Restaurateurs and Innkeepers*:
    Bakers and Millers

in *Scholars and Priests*:
    Priests

in *Scientists and Technologists*:
    Alchemists
    Chemists
    Computer Scientists
    Engineers

in *Warriors and Adventurers*:
    Soldiers

Miners and Quarriers

The roots of *mining* can be traced all the way back to the cave dwellers. These early peoples searched for both surface and subsurface stones and pillars that they could use for shelters, monuments, and weapons. At least that marks the beginning of humanity's attempt to locate, extract (where necessary), and purposefully use the rocks and minerals of the Earth for their own ends. Europe, for example, has elaborate flint mines, such as those in Cissbury and Grimes Graves, England, which are over 10,000 years old. Flint was used to make early weapons and tools. Both flint and the gem amber (fossilized resin) were mined or collected, worked, and traded on regular routes across Europe for thousands of years.

The Egyptians and Hittites both employed armies of men in the mining industry. Gold—what the Egyptians called the "body of the gods"—was a particular object of this activity, and copper was not far behind. The Egyptians mined copper and turquoise in the Sinai Peninsula as early as 3500 B.C. Bronze—copper alloyed with tin—was produced around the same time. Iron production is believed to have begun as early as 3000 B.C., although the first Egyptian records do not speak of it until the 14th century B.C. Lead remains from about 2500 B.C. have been found in the ruins of Troy. *Quarrying* was as important as mining, since rock boulders were a basic building material of the ancient Near East. The great pyramids of Egypt were built over 4500 years ago. The largest of these, Khufu, contains over two million blocks of limestone and red granite. The limestone was cut into blocks weighing as much as 30,000 pounds each on the opposite side of the Nile. These were then transported across the river.

The *miners* and *quarriers* of the ancient Near East were not free laborers. The Egyptians mined copper and

The basic material used in the construction of the pyramids pictured here at Abusir was locally quarried limestone. (Institute of Archaeology, University College, London)

turquoise in the Sinai and gold in the eastern deserts under royal direction. Indeed, since the pharaoh was the sole owner of all that was in the kingdom, he was entitled to all that the earth bore. Metals were precious commodities, and the hoarding of gold and copper placed the pharaoh in an enviable position in relation to other rulers in the ancient world. Egypt's domination of the world's copper supply was a significant factor in its world leadership during the Bronze Age, from about 4000 to 1200 B.C. In fact, the decline of Egypt's world power came at about the same time as the onset of the Iron Age, which began around 1200 B.C. Egypt had no iron, so its weaponry and tools of harvest then became outdated. Control of the mines, then, was a significant factor in world power struggles in early times, just as the control of oil fields is in modern times. A ruler was not about to leave such a critical industry in the hands of common laborers.

Slaves and *soldiers* did the mining for the Egyptians, as well as for the Hittites. Operations in the Sinai were policed by regular army troops and administered by royal officials. The *miners* and *quarriers*—the latter almost always slaves—were treated harshly and without regard for their personal health or endurance. Many miners were citizens who worked in the mines for payment of their taxes or as required civil service during certain months of the year. They were treated better than the slaves, but the working conditions were still scandalous by any standards. Yet another class of miners consisted of captives. The Sinai mines were worked extensively by Asiatic slaves. In the 15th century B.C., Canaanites were led seasonally by armed escort away from their homeland to work the turquoise and copper mines at Serabit el-Khadem in Sinai.

Miners of the ancient world were slaves, citizens, and soldiers. Their job was critical in the establishment of a country's position of authority. Hor-ur-Re, the leader of another trek to the Sinai mines, recounts both the misery and the patriotism that characterized his miners:

it was not at all the proper season for coming to this mining area.... It was difficult ... in my experience to find the [proper] skin for it, when the land was burning hot, the highland was in summer, and the mountains branded an [already] blistered skin.... my entire army returned complete; no loss had ever occurred in it.... There was no [cry of]: "Oh for a good skin [of drinking water]!"

The importance of the mines may also be detected in the fact that the viceroy for Ethiopia—one of the highest official positions in the pharaoh's court during the earlier 18th dynasty (1550-1375 B.C.)—was given three main responsibilities, one of which was the direction of operations at the Nubian gold mines, which supplied considerable offerings for Amon, the patron deity in the exploitation of the mines.

Around 1200 B.C. the Hittites began to supplant the Egyptian domination of the metal economy. They did so through massive production of the world's new metal standard—iron. Iron was a stronger and more adaptable material, suitable for the strenuous demands of war and cultivation. Its widespread use and superiority over copper and bronze sealed the fate of the once-unrivaled civilization along the Nile. Enormous stores of silver and lead were extracted annually, as is still true even today. Iron ores were taken liberally from the Anatolian mountains, and the Hittites soon developed a system of iron smelting and eventually ironworking.

Although slaves and captives doubtless worked the Hittite royal mines, a large supply of mining labor came from peasants, who apparently worked their farms during the growing seasons and the mines in the winter. They probably did not earn wages for their labors, but more likely eased their tax burdens or fulfilled civil service requirements. In any case, the system apparently became so common that the king had to make apologies for the seasonal dip in the mining and smelting industries during harvest time. Writing to one of his con-

Smelting, here of iron, was often carried on near the mine site. (From Harper's New Monthly Magazine, 19th century)

temporaries, probably the King of Assyria, the Hittite King Hattusilis III explained:

> As for the good iron which you wrote about to me, good iron is not available in my seal-house in Kizzuwatna. That it is a bad time for producing iron I have written. They will produce good iron, but as yet they will not have finished. When they have finished I shall send it to you. Today now I am dispatching an iron dagger-blade to you.

Presumably it was a "bad time for producing iron" because too many miners were still home harvesting their fields.

Mining was carried on in many parts of the ancient world. The Fifth Book of Deuteronomy in the Old Testament makes mention of a land so spectacular that "its stones are iron, and out of whose hills you can dig copper." This was presumably a reference made to the valley of Timna in Israel's Negev Desert. Copper mining was an unorganized family enterprise in the valley as early as 4000 B.C. when malachite (from which copper is extracted) was found lying loose on the desert's surface. The Egyptians, who had perfected a system of copper processing, moved into the region with their army of soldiers and miners and began to systematically exploit

MINERS AND QUARRIERS 65

the mines on a large scale for the pharaoh by about 2000 B.C. The ruins of a temple to the goddess Hathor, dating from about 1300 B.C., have been unearthed in the Negev. Constructed for the worship of Egyptian copper miners who worked the region, it included sculpted copper votive offerings and other copper objects within its chambers. The temple was later used by Midianite miners. Stone-headed hammers have been excavated from the cliffs of Timna, indicating the crude manner in which miners had to labor—smashing the sandstone walls to loosen chunks of malachite. The valley supported a bustling industry and, at least during the Egyptian working of the mines, copper-smelting furnaces burned day and night. There were apparently rotating shifts, so that the furnaces, if not the mines, could be worked as 24-hour operations.

Miners worked elsewhere, too. Archaeologists have excavated the Cave of the Treasure, dating to about 3000 B.C., in the mountains surrounding the Dead Sea. It was apparently a hiding place for metal objects, perhaps in anticipation of Egyptian invasions. Farther north, the Caucasians were also miners, and their graves are laden with rich metals and weaponry. The Urnfield peoples of Eastern Europe—so-called because they buried the cremated ashes of their kin in urns, which in turn were buried in grave fields—were pioneers in deep mining. The most famous of their communities was one established on an island in the Federsee Lake in Württemberg, Germany. The people of Federsee did heavy mining for copper deep beneath the earth's surface, and may have been the first people to employ mine shafts. By 1200 B.C. they were using elaborate systems of props to prevent wall and roof collapses. Miners built fires against the mine walls and then doused them with cold water to produce cracking for easier accessibility and extraction. *Timberers* formed an essential part of the mining team because of the props they supplied for shafts and the fuel they made available for subsurface firings as well as for smelting.

 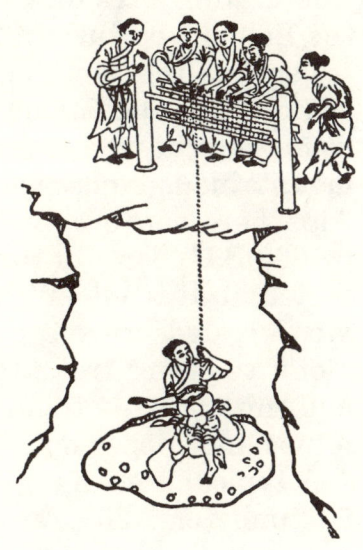

Chinese miners knew how to vent poisonous gases from coal mines (left) and lowered miners and prospectors into mines by rope (left and right). (Authors' archives)

Farther east in Asia, mining was carried on actively, too. Most of the ancient world's diamonds and rubies came from India. Central Asia was the source of the jade so prized by the Chinese. Some jade was dug out of veins in the earth, but women were often employed to pick waterworn chunks of jade out of streambeds. In China itself mining was carried on from early times. Miners were sometimes lowered by rope into pits where they dug out the desired minerals; they sometimes lit fires against a rock face and then doused it with cold water to promote cracking—a method used in many parts of the world.

Other ancient mining communities are known to have existed in a variety of places and times. One in Iran at Tal-I-Iblis, dating back to at least 4100 B.C., featured extensive copper-mining and -smelting operations. Workers there were housed in camping communities that consisted of rows of rectangular one-room housing units. The Sumerians set up mining operations much later around iron-laden meteorites. Meteorites were not a steady source around which to build an industry, though, and the Sumerians quickly depleted their sources of "the heaven metal." Other more successful iron-mining sites included Hasanlu in Iran and those of the La Tene culture in Switzerland.

Perhaps the most enlightening view of the life and work of the ancient miner and *smelter* can be obtained from Austria, where scores of copper mines dotted the mountainous landscape. At Mittersberg in the Austrian Alps there were over 30 such mines within a one-mile area by 800 B.C. Some of them ran as much as 400 feet deep, which indicated engineering genius for that time. Wooden sleds were constructed to be pushed in and out along the long treks through the mine shafts. *Porters* and miners used them to get support timbers into the mines and copper out.

A typical crew numbered about 180 men—40 miners, 60 timberers, 20 *preparers* who readied the metal for working, 30 smelters, and 30 porters and *guards*. The porters helped with general hauling and laboring, often assisting the miners directly, and the guards insured the safety of the crew as well as preventing the theft of precious metal by the crew members themselves. Miners seem to have worked as part of some civil obligation, much as today young men might be drafted into military service. They slept in temporary camps for an allotted period of time before returning to their homes in the valley and their occupations as *farmers*. Mining was apparently not a profession in the sense of workers voluntarily providing a service in return for a wage. Instead, miners served a sort of tour of duty, which harshly taxed their mental and physical energies. Since the mines were unsafe, unhealthy, and scandalously dirty, and since wives, children, and conventional life-styles had to be temporarily abandoned, it is not surprising that miners counted the days before they could return to their normal lives as fresh-air, aboveground farmers.

The Greeks and Romans continued this same tradition of imperial-style mining. Around the time of Christ the writers Strabo and Pliny the Elder offered the first known written tracts concerning the occupation of mining and quarrying, including some details of mining operations. The Mycenaeans mined silver at Laurian as early as 1400 B.C. The Greeks worked the same mines just south of Athens between 600 and 350 B.C. They built shafts as deep as 400 feet and extracted lead, zinc, and iron. Marble and potters' clay were also mined by the Greeks, who doubtlessly made extensive use of slave labor to fill the ranks of the "profession." The *Golden Fleece* that appears in Greek legends was, in truth, a sheepskin that was used by panners to net gold particles in shallow streams. The fleece was the skin, and when full it could properly be called golden. The widespread references to the Golden Fleece make it seem likely that such gold panning was rather common. It may have even been a full-time occupation for many a speculator, although its rewards were certainly unpredictable and rarely sustaining. Much of Europe's gold in this period came from the Altai Mountains of Siberia, but we do not know how the metal was mined, whether panned or dug out.

The Romans operated a wide variety of large-scale mines and quarries throughout their vast empire, includ-

These Spanish miners, with picks slung over their shoulders, are marching to work. (Authors' archives)

ing such distant lands as Britain, Gaul, Spain, and North Africa. The most noteworthy of these enterprises was the Rio Tinto mines in Spain, which yielded tremendous amounts of gold and silver, along with copper and iron. The Romans were noted less for their successful mining operations than they were for their associated engineering techniques. They employed highly trained and skilled staffs of *engineers* to help design and "blueprint" mines and shafts before they were ever opened. The methodical way that these officials approached projects was uncommon at a time when problems of wall strength and ceiling supports were typically faced only when necessary—and often too late for many a poor miner. These Roman officials—appointed by the government—were the first *mine engineers*.

The ancient miner was not truly a professional. Imperial mining monopolized the business because the whole fate of governments and kingdoms rested upon the stores of precious metals that backed monetary and trade systems, and the availability of metals for practical use—especially iron—determined the sophistication and strength of both weaponry and farm implements. As a result, the economic, commercial, military, and agricultural strength and development of a civilization depended to an enormous extent on the successful operation of its mines. The fact that they were controlled by the state or the king, then, seems only proper.

Since there were only royal or imperial mines and few private ones, the miners themselves were state workers. In fact, most of them were slaves and captives, or free citizens who traded their labor to the state in return for the liquidation of their debts or back taxes. When these sources were insufficient to fill the ranks, private citizens were drafted into state service for specified periods of time. These were usually peasant farmers of generally low birth and socioeconomic status. There was little need or interest in granting them any rights, privileges, occupational safety, or protection from grossly unhealthy working conditions. Yet the miner played a key role in the development of society.

After the fall of Rome, mining temporarily receded as a major economic activity in Europe. In succeeding centuries, especially after the rise of the Islamic world, some of the most active mines were in northwest Africa. The two most-prized items were gold and salt. The Sudanese, in the region of the Niger River, were so desperate for salt that they sometimes traded gold for it, ounce for ounce. The salt mines were in the central Sahara, in a wasteland so harsh that the slaves who were forced to operate the mines had to rely on heavily salted water from the mines for drinking—and would starve to death if their Arab overlords failed to arrive on time with the necessary supply of food.

How and even where the Sudanese dug their gold we do not know, for they kept secret the site of their gold diggings. Even the trading was done long distance, in a form of silent barter. The Arabs would lay their goods in a designated spot and retire; only then would the Sudanese emerge from their "holes" (presumably their mines). They would set out an amount of gold that they thought equaled the value of the goods and would leave the scene. If the Arabs agreed, they took the gold and the trade was concluded; otherwise the piles were adjusted until both sides were satisfied. Arabs, and later, Europeans, tried for centuries to locate the Sudanese gold mines, but without success; and by modern times those mines were largely played out, though others farther south gave rise to the name "Gold Coast."

Medieval Times

Meanwhile, in medieval Europe, decentralization of government meant that mining operations became more spontaneous and sporadic, generally lacking in both organization and productivity. During this period of rela-

This well-dressed miner worked in the tunnels, shafts, and quarries of late 16th-century Europe. (By Jost Amman, from The Book of Trades*)*

tive neglect there was a gradual erosion of the previously absolute authority of the existing imperial powers over the exploration, operation, and profits of the mines. Mining became more of a chosen occupation and less of an unpaid obligation. By 1000 A.D., mining was established throughout Europe as an essential industry. Miners were sometimes private *prospectors* and hired *diggers*, but at other times the term referred to the promoters and financial underwriters of mining or quarrying operations. The latter group included noblemen, wealthy merchants, and even kings.

There is frequent mention of many thriving mines by the 11th century. Tacitus may have written in 98 A.D. that "the gods have denied them [the Germans] gold and silver, whether in mercy or in wrath I find it hard to say: not that I would assert that Germany has no veins bearing gold or silver: for who has explored there?" But as early as 968 A.D. silver was being mined in Rammelsberg near the rich German copper-mining town of Goslar in the Harz Mountains. The mountain range was soon bustling with the activity of miners extracting precious metals from the previously unexplored veins. In

1136 A.D. there was even a silver rush in the German region of Frieberg, when *salt traders* discovered a fine-grade silver ore there. Many prospectors and adventurers tore across the countryside with picks and shovels to seek their fortunes. Thirty years later many had found success, and a sprawling mining and smelting industry provided ample employment for miners and *metalsmiths* in a factory town of over 30,000 residents.

While the mines of Central Europe were well known for their gold and silver, France had the greatest stone quarries in the world. Stone was a common building material during the Middle Ages, used for both homes and public buildings. The underground quarries from

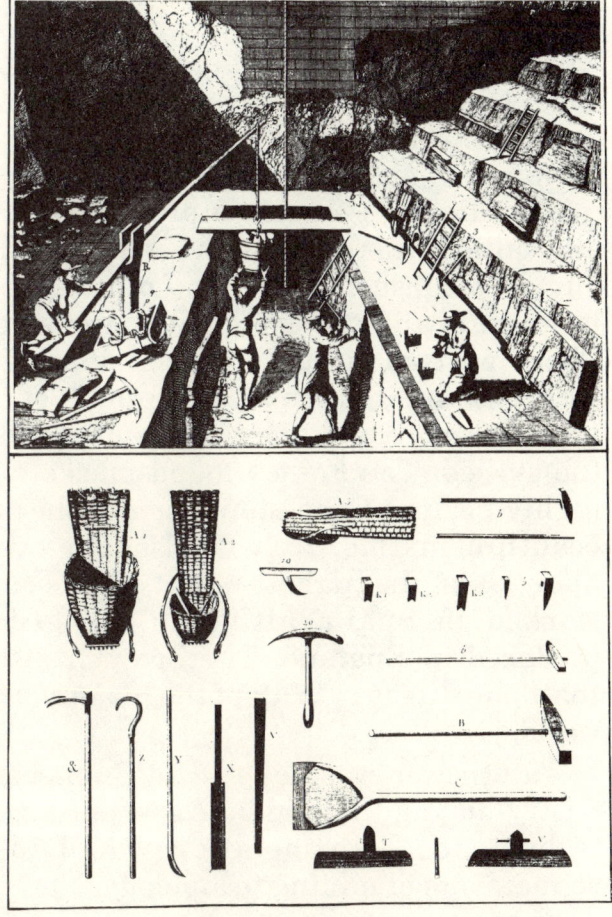

Slate was mined from well-planned quarries like this one at Anjou. (From Diderot's Encyclopedia, *late 18th century)*

MINERS AND QUARRIERS 73

which it was mined were elaborate mazes of tunnels that were excavated throughout the land. So extensive are these systems that many French communities—greatest among them the city of Paris—are widely known as "hanging towns." Quarrying at the famous mines of St.-Leu d'Esserent was carried on in shafts that extended over a mile from the entrance. These mines were worked continuously for eight centuries before being severely damaged by World War II air raids. Quarriers often worked the mines in layers, building one system of tunnels beneath another. There were as many as three levels of workings in such cases, and certain areas were sectioned off for workshops in which quarriers chiseled the stone into blocks.

Iron ore was another important object of the medieval miners' labors. It was mined extensively in England, where every town had its own ironsmith. The Forest of Dean had such a developed system of heavy iron mining and ironworkings that it has frequently been called "the Birmingham of the Middle Ages."

The Middle Ages was a prosperous period for miners and mine owners. Stone and metals from both open air and deep subsurface pits and mines were extracted on a large scale and in a methodical, well-organized manner. There were many uses for the stone and metals, uses which stimulated the steady growth of the industry and provided more and more employment for the hardy and the skilled. The French found markets abroad for their highly desired Caen stone, as did the Italians for their beautiful marble. Iron was sought everywhere for its many uses; the highest-quality iron seems to have been worked in Spain, although the greatest bulk was produced in England. The Spanish also produced steel that found its way to all of the major ports of the Western world.

Waterpower was adapted to the needs of the iron industry during the Middle Ages, permitting new heights to be reached in metallurgy and inspiring greater refinements of mining technology. New ideas spread fast throughout the Western world for many reasons. One was the great repute enjoyed by German miners. Because they were considered the greatest in the world, their services and instruction were sought by many a foreign king and mine owner who wished to make the most of a new mining enterprise. In the 12th century the kings of Transylvania made a habit of calling in German miners to assist in their operations, and in the following century the Serbian rulers did the same. In 1303 a four-man team of German miners was requested by the English Crown to prospect mining regions in the Flint of Wales for rich minerals. The high esteem in which they were held may be indicated by the fact that they were each salaried at seven shillings, six pennies a day, while their English counterparts were paid only two to three pennies. For all their expert knowledge and skill, however, this particular team presented the crown with supposed copper ore findings that turned out instead to be worthless copper pyrites. Nonetheless, German miners were employed all over the world and spread their knowledge of engineering, mining, and metallurgy throughout Europe.

The Germans were not alone in spreading the knowledge of mining technique and theory. Many miners from different parts of the world were wanderers who carried with them new ideas. The Cistercian

Quarriers are shaping slate blocks, while a cart prepares to haul a load away. (By W.H. Pyne, from Picturesque Views of Rural Occupations in Early Nineteenth-Century England*)*

monks—also known as the White Monks for the color of their habits—helped disperse technical innovations beyond the borders of kingdoms and petty feudal states and manors. They were well known for their high-quality work and updated standards in both agriculture and industry. Every monastery had its own factory, which was usually right beside the church. The latest techniques of waterpower were used to run the most-modern machines known to Western Europe.

At the base of the extensive Cistercian industries was the mining operation. Iron was the most commonly worked ore because of the many uses that the Cistercians found for it within their industrial complexes. Iron ore deposits as well as forges were frequently given to them as donations. Other mines were taken over by the Cistercians as they expanded their operations first in France and eventually throughout Europe. The Abbey of Clairvaux absorbed a great many mines in its mineral-rich Clairvaux region, and between the 13th and 17th centuries the White Monks were the top iron producers in the entire region of Champagne. They also mined lead and copper to a lesser extent, and even a bit of silver and gold came under the blows of their picks.

New mining techniques and engineering feats enabled a more systematic and economical extraction of minerals and stone than ever before. Wandering miners and the broad influence of the Cistercians helped to spread these innovations all over the Western world. During the early Middle Ages the Carolingian emperors had made silver the foundation of their monetary system. Before long there was a wild hunger for silver, and new large-scale mining techniques were invented that benefited the mining industry as a whole. By the 13th century the gold standard was reestablished as the basis of European monetary systems. The combination of these factors, plus the ever-present exchange of ideas via colonization and petty warfare, made for a flourishing and constantly improving medieval mining industry.

Throughout medieval and early modern times, the

term *miner* was confusingly used to refer to both the digger in and the owner of the mining operation.

Even among mine laborers—the people we today commonly regard as *miners*—greater freedom and professional opportunity existed in the Middle Ages than had been possible in the ancient imperial and royal mines. Private mine owners, even medieval royalty, did not have the vast human resources available to them that the ancient kings and emperors had always seemed to possess. Slavery was uncommon and peasants owed allegiance to their lords, who typically demanded agricultural labor, with little surplus time that might be spent on mining. The Middle Ages was a period characterized by struggles for power between kings and aristocrats—struggles that had rarely surfaced in the great, ancient empires. More often than not, the power of medieval kings was greatly curtailed by the interests of the noble lords.

Miners and prospectors were given great leniency by landed lords and kings, who were eager to obtain a portion of the wealth to be had from successful mining enterprises. Many common people complained bitterly of the miners' right to prospect virtually anywhere they wished, including churchyards, highways, and the good land of many a farmer. To gain some control over these practices, the Central European overlords set up a mining administration, headed by a *Bergmeister*, who was the prototype of the English *bermaster* or *barmaster*. Such administrations served not the general public, but rather the vested interests of the miners, who sought prosperous ventures, and the overlords, who sought to reap a percentage of the intake. Miners were granted land to set up camp on a prospective site and, if the venture proved successful, a *mine city* was formally incorporated; all its residents were granted the none-too-common status of freemen—free, that is, from the ordinary shackles of feudal obligation. The people who actually worked the mines in such a city were additionally allotted free ale and baked items, free transportation of goods, freedom from guild regulations, and exemption from military service.

English miners had enjoyed a similarly privileged status as early as the Bronze Age, when the famous *stannaries* (tin mines) of Devon and Cornwall attracted free miners. Indeed, these stannaries continued to be worked into the Middle Ages, although many of the free miners migrated to Germany, where they directly influenced mining codes. The high status of miners was condemned by many a manorial lord, who felt the repercussions of bonded peasants leaving the fields to be miners and, subsequently, free men. But English kings were so tantalized by the wealth of these stannaries that they routinely approved and increased miners' privileges.

But despite their liberal privileges, miners also suffered from serious exploitation. To begin with, unless the minerals and stone being extracted were extremely valuable, miners were openly ill-treated and misused. Silver and gold miners in Central Europe and tin miners in England may have enjoyed liberal freedoms and privileges, but miners of the more-common and less-precious metals were unlikely to be courted so sweetly. They received considerably less pay and were burdened with countless restrictions on their prospecting activities. Regardless of the social esteem or financial status of different classifications of miners, all suffered from common problems inherent in the industry and the economic structure of society.

The labor of miners was exhausting and demanding. They had to haul heavy cartloads of lumber, rock, and minerals up and down long, steep, dirty, treacherous shafts. They worked in all weather conditions and put in extremely long hours, sometimes even into the night while lanterns lighted their way. Often miners were given additional duties as *watchmen* on the all-night shift in order to protect the mine's precious take. Sometimes they assisted in various phases of working the metals, and quarriers usually played various roles in the rough-hewing of boulders, which took place in underground workshops. Miners also assisted the timberers in the placement of props to build up and support galleries (the tunnels through which miners passed to reach different parts of the mine) and workshop walls. In the quarries, stone pillars were often constructed by miners for the same purpose. Often miners even had to train for and perform duties as *soldiers* to protect the mines from *robbers*.

Of course the amount of hard labor miners had to do depended largely on the value of the product being mined. Silver and lead miners were considered experts in their field, highly skilled and technologically knowledgeable; they could depend to a great extent on the assistance of porters and apprentices. The quarriers and lesser miners, on the other hand, were regarded as little more than common laborers. They could expect no help, but were stuck with all the chores, both those demanding skill and knowledge and also the routine *dirty work*—literally, the digging of dirt.

Mining continued to be an essential industry through the Renaissance, Reformation, and into the early modern

Miners often used dangerous underground fires and explosions to break loose the rock they wanted to extract. (From Diderot's Encyclopedia, *late 18th century)*

era. One of the most notable works depicting the industry of metallurgy and the lot of the miners was by Georgius Agricola's classic volume *De Re Metallica*. One of the most interesting observations made by Agricola was that the mining profession was commonly maligned and suffered low social regard because for so many centuries it had been the occupation of slaves and criminals sentenced to work the mines. In the years before the Industrial Revolution, mining underwent some modest changes. Most notably, waterwheels—used to pump out water in Germany, Bohemia, and Hungary since the 15th century—became more widespread. The same energy was later put to use in hauling boxcars up mine shafts. Wooden tracks already had been used on the Continent for this process. These modernizations of European mines resulted in the reclamation of many operations that had been shut down for years. Until the 19th century, when the Industrial Revolution had a tremendous impact on the profession, labor and work conditions remained basically unchanged. Miners continued to be an important segment of the laboring class, one hideously exploited by mine owners.

The Industrial Revolution

The most significant change in the mining profession came about with the advent of the Industrial Revolution and its accompanying demands for mechanization and energy. Machines were being invented to ease the burden of human labor. Raw and processed metals were needed to make the machines, and energy was necessary for their operation. Iron ore became the metal in greatest demand and coal the chief fuel.

Before the Industrial Revolution, the most popular fuel had been wood, often in the form of charcoal. Coote's ironworks in Ireland employed over 2,000 men doing nothing more than preparing wood for fuel. This activity conflicted seriously with the interests of the lumbering and shipbuilding industries. To help ease the situation, the British Parliament limited licensing for new furnace construction and ordered that all trees cut be replaced with seedlings. With the depletion of forest reserves because of extensive and irresponsible woodburning, coal became the most important fuel.

Coal had been used in English iron pits since the 13th century, but was not appreciated for a long time, there being little pride or profit in being a coal miner. In Staffordshire coal could be dug from the surface or from very shallow mines. But it was so plentiful that it was blatantly wasted. In 1660 two million tons were mined; of that total, five thousand tons a year were senselessly discarded as useless slack.

It was long thought that coke (an extract from coal) was too impure and could interfere with the metallurgical process. Outside of England coal was even less understood, and a 1781 French treatise on commerce matter-of-factly concluded that coal could not be burned.

Only in China was coal's value recognized and exploited early. By the 11th century, the Chinese had replaced charcoal with coal for metallurgical uses; they also pioneered in the use of explosives to aid mining, long before gunpowder was even known in the West.

Quarried rock was often ground to appropriate size for road surfacing right on site. (From Centennial Exposition, *19th century)*

It was England, where industrialization would restructure the foundations of modern society, that paved the way for great opportunities and technological advances in the mining industry. As the leading coal-mining nation in the world—greatly outdistancing its nearest rivals—England was the locus of the profession's growth during the pivotal era of the mid-18th to mid-19th century.

Because it involved so many laborers working together, the mining industry would become the testing ground for a great many new ideas in industry, from the perspectives of both labor and management. It would also attract some of the earliest attempts at mechanization and reform of wage and working conditions, because it was so highly labor-intensive and because it represented the base of the Industrial Revolution. Without coal, there could be no factories for mass production. Without steel and iron—industries that themselves needed coal for sustenance—there would not be such essential items as rails for the transportation system that increased communications and permitted the Commercial Revolution, which followed on the heels of the industrial one.

Miners in England were mostly employed in the extraction of coal by the 18th century. Laborers faced extreme drudgery and danger to earn an honest wage. Individual deaths occurred almost daily, and frequently there were massive disasters. Death came mostly from sudden flooding or suffocation, when not enough oxygen existed in the mixture of damp and dust that permeated the underground pits. *Chokedamp* was the term miners used for the suffocating coal dust that mixed with the ever-present damp rising from the mucky mine floors, with their constant streams of seepage. Trapdoors were used to let in fresh air and to release the chokedamp from the shaft.

Young boys called *trappers* were usually employed to operate the trapdoors. They had to constantly open the one, while closing the other, alternating the good and bad air. If this was not done right, or not frequently enough, or if the poor lad (who himself often worked over 80 hours a week) was to fall into a desperate moment of dozing, the miners would feel the choke overcoming them. In their terror they might first begin to scream at the so-and-so who was minding the vital trapdoors and then—if there was still no sudden rush of fresh air—they would scramble to get out of the mine as quickly as they could. If the trapper could rectify the situation in time, or if the mine shaft was not too deep or treacherous, they might make it to safety. But often they did not. Then the mine had to be "aired" for a short period of time, while new laborers were gathered to replace the deceased. The mine was worked again in a short time.

Floods also posed a constant danger, especially in the deeper pits. As veins—those earthen regions being tapped for their rich deposits—were worked progressively, new water sources were constantly finding their way into the shaft. It was bad enough that miners had to work in sloppy and damp conditions, with water levels often rising above their knees, but occasionally a main spring was tapped that allowed a torrent of floodwaters to kill almost instantly whatever crew was there. Other times, support walls or beams suddenly gave way to the pressure of the surrounding waters with the same result. Women and boys were hired routinely to haul out bucket after bucket of seepage buildup, but their backbreaking efforts were of no help in times of sudden flooding.

Firedamp was yet another significant safety hazard. As mine walls were constantly hacked away, swirls of methane gas were released into the atmosphere. Although much of it was released through the trappers' ventilation efforts, a considerable portion of it became entrapped in air pockets that lurked invisibly in the mine's atmosphere. Mines were always dark, of course, being deep under the ground. In order to carry out their work, workers carried miners' lamps, which were open fire lanterns. The problem with this practice was that occasionally some poor fellow busy at the work of earning a day's wage unwittingly moved his lantern into one of the methane pockets. It immediately exploded, certainly killing the miner carrying the lamp, and sometimes even others in the immediate vicinity. Many a miner was blown to bits in this way, and some mines had such a high incidence of these tragedies that they were forced to close down because they were deemed "too fiery" for normal operations. Even large quantities of concentrated coal dust sometimes reached the point of combustibility.

Even when miners were not being killed, their working conditions were deplorable. Besides the exhausting task of hacking into pit walls of stone, there was the enormous difficulty of moving support timbers into the mine and of transporting the mineral or rock takes out of it. Even the transport of workers from above ground into the depths of the pits could be a treacherous ordeal. With lifelines of hemp, miners were lowered sometimes hundreds of feet straight down into coal pits, where they might begin their work. A hemp *windlass*—a device designed for hauling men and materials up and down the shafts—was fastened around the miner's body; he was careful to keep his legs probing while he was being lowered, lest he bash his head against a protruding boulder. The haul up and down was a long one, and sometimes the hemp just was not up to it, especially after having rubbed against shaft walls and dry rock countless other times. Women and boys, with the help of horses, usually worked the windlass from above ground, but few falling deaths were caused by their negligence or error. They knew how critical a chore they had, and they usually did it with great care. But hemp was expensive, especially the lengths needed for deep mining operations. Windlasses were infrequently inspected and were replaced voluntarily by many mining companies only if they were obviously ripped or shredding. Even when they finally broke, they were frequently only repaired rather than replaced. A widow or son was usually permitted to replace her late husband after a fatal accident. She could, however, expect to be ridiculed and scorned because of her immodest dress due to the excessive heat underground, and the masculine appearance that she soon attained from deep coal scars and a general darkening of the skin.

Because of the cost of hemp, windlasses were sometimes abandoned altogether in favor of ladders. In the deeper mines, ladders could not reach all the way from the ground level to the bottom of the pit. In such cases a series of ladders were used, each one arriving at a narrow ledge, which could be used as a landing from which the next ladder could be reached. As if such a climb—going up hundreds of feet—were not difficult enough, the miners even had to haul cargo on their backs at the same time: tools, timbers, water, and even the coal itself. Sometimes a mine had a tunnel that could be used for hauling the coal gradually to the surface. Such hauls were long and burdensome, especially since the shafts were only barely large enough for a woman or child to stoop in, and often the haulers could only crawl along the muddy floor. Flat baskets filled with coal were tied to their backs in preparation for the haul. They had to stoop severely or crawl in order to keep the coal from falling out of the open baskets. Sometimes small carts or wagons were dragged and pulled, but the burden remained overbearing and many a frail woman or young lad would

Women often worked in coal pits hauling coal out of the diggings. (Authors' archives, 1842)

eventually succumb to exhaustion, thereby inviting the wrath and condemnation of their foremen and supervisors. Sometimes they even lost their jobs for repeated failure to hold up for a good 12-hour shift. Strong men may have been better suited for the job of hauling, but they were needed more in the pits. Besides, mining companies did not care to foot the expense of having to build shafts large enough for a full-grown man to move through.

Miners also still faced the constant health hazard of breathing in coal dust and silica and of standing up to overtaxing burdens of load and labor. One of the more common of miners diseases was silicosis, caused by the inhalation of free-formed silica (quartz). Nonetheless, little regard was ever given to health precautions.

In North America, mining did not develop so quickly. This was only partly due to restrictions that Great Britain placed on the industry in order to force the Colonies to import manufactured goods from the British. Many of the restrictions were placed not so much on mining itself but on the manufacture of goods that would require mined metals and fuels. The early growth of the profession was also slowed by the relative lack of accessible water routes to transport bulk rock and mineral takes. Although the North American continent has a long coastline, most of the early mining was done in inland regions such as the Appalachian Mountains, especially in Pennsylvania. The significance of being able to establish mines near water routes cannot be overemphasized, since water transportation was the chief means of moving products to markets. Even light items were difficult to transport over land, but heavy ones like coal, lead, and iron ore were virtually out of the question.

Even so, Colonial America saw some mining activity, although the industry was a long way from the world-leading position it would assume in the 19th century. There was a critical labor shortage in the early Colonial era, and Native Americans were used to take up the slack, a practice that was especially common in the copper mines in the Lake Superior region. A more typical solution was to purchase or rent slave labor. In 1750 one of the earliest commercial coal-mining companies was using Black slaves in its operation near Richmond, Virginia. During the American War of Independence, slaves were used to dig the coal required to fuel ammunition factories for the cause of independence. The first mining company in the Colonies was "the Company of Undertakers for the Iron Works," which operated in the Massachusetts Bay Colony between 1643 and 1677, when it finally succumbed to bankruptcy. During those years it attracted a fair number of skilled miners and ironworkers from England and Germany. Coal and iron were mined in the Appalachians from very early Colonial times, and by 1780 there were established ironworks in all 13 states. Lead and zinc were first mined in Missouri in 1718 by the French. Lead was later to be used by the Continental army for ammunition in the American War of Independence.

But it was in the 19th century when the newly independent United States saw a revolution in its mining industry. Copper was first mined in Michigan in 1845, the same year that iron ore was first extracted from fields in the Lake Superior district. Gold was discovered in 1848 in California and later in Alaska, and the famous gold rushes followed. Thousands of people from the East became overnight prospectors, leaving their homes in covered wagons, bound through thick forests and across dry deserts to the golden West. Some of them struck instant wealth and good fortune for their pit-mining or stream-panning efforts. But even more spent endless hours, months, and even years, cramped over a stream with a sieve-like pan, hoping vainly to capture enough gold particles so that they would never again have to work for a wage. Many prospectors were ultimately forced to hire themselves out as miners in order to earn a living. Mining towns sprang up throughout the West. Silver and copper were mined in Nevada, gold in Colorado and the Dakotas, and copper in Arizona and

Nevada. Trails for teams of covered wagons were blazed, and stagecoach and post routes, like that taken by the Pony Express, were developed to maintain communications between the established East and the new mining cities founded on the western end of the vast nation. The mining profession—and particularly gold prospecting—played a highly significant role in opening up the American frontier.

The 19th century was one of maturity for the mining industry and its related occupations. New inventions and innovations made the process of mining somewhat easier and safer, and therefore more economical. Coal miners became better organized, and by the end of the century had formed labor unions to pursue workers' rights and reasonable protection. Legislation was enacted to make mines safer to work in, to limit exploitation of labor by mine owners and operators, and to regulate the wages of those who worked the pits. In fact, miners formed one of the strongest labor coalitions in both England and America, and were directly responsible for the initiation of unprecedented labor cohesiveness and governmental recognition of both the rights and the power of organized labor. The entire labor movement was focused on the most-important faction of the mining population—the coal miners.

As late as 1790 Ben Franklin—wanting the new nation to establish a paper currency—argued that "gold and silver are not the products of North America, which has no mines." Of course, six decades later, gold and silver were to become very much the products of North America. But even before that—and with a significance surpassing that of the more glittering riches—coal was ingrained in the heart and soul of the history of American mining. It was coal that would fuel the new industries that provided the materials for the nation's growth, and it was the coal miners who fought tirelessly to establish the human dignity and rights of American laborers in all industries.

Coal first became commercially significant in America in the early 19th century when Pennsylvania—during the heat of the Industrial Revolution—experienced a severe wood, and therefore charcoal, shortage. By 1820, Pennsylvanian mines were shipping 365 tons of coal per year; 10 years later that figure rose sharply to 360,000. In 1846 a Canadian researcher, Dr. Abraham Gesner, first extracted *coal oil* (kerosene) from coal. Kerosene, used to fuel lamps in the days before electricity, came into great demand, and coal miners were urged to increase their efforts even more. In 1856 Henry Bessemer invented a converter that made possible a cheap steelmaking process. More coal was needed to fuel that popular undertaking. By 1873, 22 million tons of coal were handled by some 50,000 employees—both miners and general laborers. This made coal mining one of the leading occupations in the country.

But working conditions in America were no better than in England. For a long time the industry relied heavily on slave labor. This was expensive and, since a great deal of the major mining was done in the North, where the abolition of slavery was becoming a fervent social issue, new labor markets had to be tapped. At about this same

Horses and mules were often used to haul coal wagons on underground tracks. (From Harper's New Monthly Magazine, *19th century)*

time Scotch, Irish, and German immigration was rapidly increasing. The mine owners were quick to lure the new laborers under the ground with promises of high wages. Of course, part of the reason they wanted immigrant labor was to avoid the high wages demanded by native-born workers.

Meanwhile, coal miners in England were finally having some of their labor reform demands met by Parliament. Hobhouse's Act, a piece of legislation directed at cotton mills only, had been considered rather liberal for the times: in 1831 it reduced the workweek of children (those under 18 years of age) from 72 hours to 69, while prohibiting them from performing night work. The mines were not covered by such restrictions, and children commonly worked over 80 hours a week, including all-night shifts. (In truth, being underground that long must have made it seem hardly worthwhile differentiating between night and day.)

Lord Ashley headed a commission designed to study labor conditions specifically in the coal mines. Horrified by what he discovered, Ashley waged an all-out campaign to aid miners in their struggle for human decency and compassion. In 1842 he effectively legislated a bill prohibiting all women and children under age 10 from working underground. The bill also provided for the appointment of official inspectors to enforce the law and seek to rectify other abuses. Two years later Ashley proposed an across-the-board 10-hour workday; this idea was too far ahead of its time, however, and was finally defeated.

Later in the same year, the determined Ashley pushed through another act that declared women to be "protected persons" in the mines; further reduced the workday for children under age 13; provided that regular times be allotted for meals; forbade work after 4:30 P.M. on Saturdays; increased fines for violations; and forced mine operators to check more carefully into workers' false certificates of age. In 1847 Fielden's Act was approved, limiting all "young persons" (a legal term defining children between the ages of 13 and 18) and women to a 10-hour workday. Other Factory Acts continued to be issued throughout the century, slowly but steadily improving working conditions.

Mine operators in England and North America were happy to oblige workers on some of their demands for safety and especially labor-aid mechanization. Several of their safety precautions would permit otherwise unworkable mines to be kept open. Such was the case in 1815, when Davy's lamp was invented to provide increased worker protection from firedamp methane explosions. The use of the lamp led to the reopening of many "fiery" mines that had been forced to shut down earlier in the century when a rash of disastrous explosions sent shudders through even the grittiest of the underground hackers.

Improvement in both mining methods and machinery permitted safer digging in the pits. Deeper seams of rich coal could be tapped and even new fields could be opened. The steam engine began methodically removing water from the pit floors in the early 1800's, thus leading to deeper and safer digging. By 1840 powerful fans had been developed to expel explosive methane along with the deadly coal dust and chokedamp. Some mechanical aids—like Willie Brown's *iron man*, a rapidly pounding pick developed in 1761—were introduced. But they would not have a major impact before the 20th century.

Still, by the mid-1800's, engines, iron, and steel were making the work of the miner more tolerable than had ever been dreamed possible. Engines were used to haul loads of heavy timber, coal, rocks, and minerals. Steel rails were beginning to replace wooden ones, making hauling an even smoother task. The efficient mechanization of hauling was especially welcomed in England, after women and boys were barred from such underground work. In fact, some observers felt that it was labor laws like those that Ashley delivered to Parliament that finally forced operators to improve working conditions in the name of efficiency and labor management, where pleas for humanitarian regard had failed to attract sympathy in the past. In any case, the workers were just happy to see things getting better, a feeling that this popular mine song of the period expresses:

> God bless the man wi' peace and plenty
> That first invented metal plates
> Draw out his days to five times twenty
> Then slide him through the holy gates.

By the mid-19th century the long-exploited European mines were declining. Continental mining never had rivaled the British coal-mining operations, if only because there were few good water outlets for the transportation of heavy cargo. Gold was discovered in Australia, Alaska, and the Canadian Shield all at about the same time as it was in California. An Australian gold rush began in 1851 that was very similar to that in the United States two years earlier. Toward the end of the century, copper deposits were found in Rhodesia (present-day Zimbabwe) and the Belgian Congo, but the greatest African mining operations were opened in South Africa, which today still boasts the world's largest single supply of diamonds and gold. Many wars throughout history have been fought over mining rights and the use of native slave labor in the mines.

Modern Times

The modern status of the mining profession has largely been shaped more by the American coal miner. The strides made by British miners in terms of labor reform raised the question of whether *laissez-faire* policies should be challenged when they led to corrupt, abusive, and inhumane practices by business owners. America, too, was based on a political system that believed in noninterference of government in private enterprise, especially since independence from British interference had only recently been won. When American miners first tried to organize labor unions, they were labeled as socialistic, unpatriotic, and greedy.

Convict, Black, and immigrant labor was courted by the large mining companies in order to create dissension among workers who had different cultural backgrounds and even different languages. Sometimes minority groups of workers were even used to break strikes by crossing picket lines. Convict labor was rented cheaply and, since prisoners could not possibly organize into unions, they represented a popular supply of labor after the Civil War. Corporations made a policy of luring immigrants into the service of the mines, and in the 1870's there was a large influx of Italian and Slavic miners into the nation. They worked much more cheaply than the already-established Americans, were often unskilled, and in these early days took little part in the labor movement. Blacks, too, were considered "safe" to hire. They had only recently won freedom from slavery and were not about to organize unions, especially since only a few had any education. The director of one mining corporation declared flatly in 1877: "With the mines filled with colored men . . . the company will not be burdened with the expense of another strike for many years."

Low wages were upheld intellectually as Christian safeguards against the sins of leisure and luxury. One of the early labor reform organizations—the Molly Maguires of Schuylkill County, Pennsylvania—was violently opposed, and many of its leaders were roughly "tried," convicted, and executed in 1876 and 1877. Franklin B. Gowen, the president of the main coal company in the area, later admitted that "the name of Molly Maguire being attached to a man's name is sufficient to hang him." Yet the fight went on, and in the early 20th century John L. Lewis—himself a miner—championed the cause of the miners. As a young lad in the mines, Lewis had once defended himself against a raging mule. After he killed it with a pipe, he carefully impacted its wounds with clay and declared that the beast had died of a heart attack. Otherwise he knew he could have lost his job or worse. The incident certainly supports the later contention of one labor official that "they [mining corporations] value a mule more highly than a human being."

Miners in North America had a bitter lot, even into the 20th century. Mechanization and labor reform somewhat bettered their conditions, but mines were still unhealthy and unsafe. Most miners had chronically raised pupils, the result of constantly watching the ceilings above their heads for falling rock and slate that often proved fatal. A list of Molly Maguire fugitives posted by Pinkerton's Detective Agency in 1879 told of a William Love who had "a scar from burn on left side of neck under chin, and coal marks on hands"; and John Flynn who was "round shouldered and almost humpbacked"; and Thomas O'Neill: "Stoop shouldered; walks with a kind of a jerk"; and Patrick B. Gallagher who, besides a "large coal cut across the temple" also "wears a frown on his countenance."

Miners had good reason to frown. Between 1910 and 1945, some 68,842 were killed while working and another 2,275,000 were injured. For years they were cheated out of even claiming proper insurance coverage for their misfortunes. One labor leader summarized this scandal:

> When a workman was injured in the shop, mine or on the railroad, the claim agent...would...present himself with an instrument of agreement for the injured man... to sign. By the terms of this instrument, the company would be released from all responsibility.

And as if things were not hard enough for the men, the children too were the victims of the evils of such exploitation. Stephen Crane once wrote about "the little slate-pickers" who were employed to separate slate bits from pure coal. He described them as "a terrifically dirty little

Before the days of child labor laws, the youth of many a young boy was wasted in the mines. (By Lewis Hine, National Archives, Records of the Children's Bureau, 102-LH-1941, Pittston, Pa., 1911)

band" that "seemed proud of their kind of villainy." This description continues:

> Through their ragged shirts we could get occasional glimpses of shoulders black as stoves . . . . they live in a place of infernal dins. The crash and thunder of the machinery is like the roar of an immense cataract. The room shrieks and blares and bellows. Clouds of dust blur the air until the windows shine pallidly afar off. All the structure is a-tremble from the heavy sweep and circle of the ponderous mechanism. Down in the midst of it sit these tiny urchins, where they earn fifty-five cents a day each. They breathe this atmosphere until their lungs grow heavy and sick with it.

John L. Lewis, as president of the United Mine Workers for many long, productive years, fought passionately to organize mine labor and to create a system of *closed shops*. Closed shops were unions in which all employees of a particular company must be members, as opposed to *open shops*, which gave the laborer a choice of joining or not—thereby leaving him open to company manipulation and intimidation. On the eve of World War II, Lewis won a federal appeal to the U.S. Conciliation Service to allow closed union shops in the captive mines.

Lewis led a series of other strikes in 1946, insisting that

miners ought to have a voice in wage negotiations and in the formulation of health policy. He even proposed a royalty to be paid on each ton of coal mined to finance health and welfare programs for miners. World War II had just ended, and coal contracts were down. Increased mechanization took away jobs from many mine laborers.

Yet, the struggle for better conditions could not be slowed down. As Lewis saw it, even though mechanization stole jobs, it ought to be supported: "It's better to have half a million men working in the industry at good wages, high standards of living, than it is to have a million men working in the industry in poverty and degradation." The 20th century has seen the gradual improvement of health, safety, and organizational legislation to protect miners throughout the world.

Mechanization sped up the mining process, and many formerly bustling mine areas were quickly exploited and then abandoned. The Appalachian region of the United States is a sorry case in point. Yet the same machinery that took jobs and closed mines has resulted in much more efficient means of extracting the Earth's elements. In 1900 the average U.S. coal miner produced three tons of coal per day. Those figures increased to 4.5 by 1925, 6.3 by 1945, and 13.7 by 1964. Strip-mining, which accounted for nearly one-third of all the mining in the United States by 1964, retrieved 29.3 tons per man-day in the same year.

Many machines have contributed to increased productivity. Early in the 20th century, coal-cutting machines began to be used in West Virginia and Illinois. Everywhere electric engines were being used to haul loads and to run conveyor belts and mechanical shovels. Electric locomotive haulage began in 1924, two years after mechanical loading had been made possible. In 1936 the Russians began hydraulic mining to make accessible their unusually steep-pitched seams. Strip-mining was well established by World War II, and the continuous mining machine with its huge rotary cutters became a commercial entity in 1948.

Other forms of mining have become extremely important in the 20th century as new materials have gained in significance. Strip-mining and surface mining, while virtually destroying the ecological balance of a region, have been used more frequently than underground mining, because they result in higher yields and fewer safety hazards. Deep-sea mining has recently gained a considerable amount of attention, because of the hope of extracting rich manganese nodules (knoblike lumps) and because the vast majority of the Earth's surface is beneath water. Mining on other planets is being seriously considered, too, as the world's population demands more raw materials than it can apparently produce or discover on Earth.

Oil, natural gas, bauxite, and uranium are among the more-important underground riches that were not exploited until the 20th century, but that now provide new and sometimes unique types of jobs for miners and *drillers.* Bauxite—essential in the manufacture of aluminum, an important metal in the 20th century—has become a widely sought element that has attracted a new wave of prospecting and mining. Some otherwise underdeveloped countries such as Jamaica and Surinam have recently become significant suppliers of bauxite.

Radioactive uranium, too, has been significant only in recent years. It is used in nuclear reactors and, as such, is extremely important and in demand, particularly by the world powers. South Africa extracts uranium as a by-product of gold mining in the Witwatersrand area near Johannesburg. Prospectors seek uranium with "Geiger" or "scintillation" counters, both used to detect radiation emissions. Many efforts at mining uranium are met with violent popular opposition, because of the fear that high levels of radiation are released into the atmosphere during the process. In the state of New Jersey, popular objection resulted in the denial of such mining rights being granted to major petroleum interests.

Oil and natural gas mining require more specialized drilling techniques. Their importance as prime energy sources in the highly industrialized nations of the world is unparalleled. (Occupations in the oil business are discussed later in this book in the article "Well Diggers and Drillers.")

Mining today can be a very dangerous profession, when dealing with deadly materials such as uranium, or when using treacherous techniques, such as those used in the extraction of oil and natural gas, particularly in continental shelves or even in the deeper seas. Mining is a major employer of people throughout the world, and the profession is more refined and technical today than ever before. Besides laborers and machine operators, mine crews usually include university-trained *geological* or *exploration engineers*, who locate and evaluate mineral deposits; *production engineers*, who consider the most appropriate and economical techniques of extraction; and *metallurgical engineers*, who oversee the smelting and refining of the raw metals, oils, gases, and minerals.

Perhaps the most significant social effects of the mining industry today lie in the fact that there is a greater demand than ever to uncover the Earth's wealth—which is rapidly vanishing and increasingly difficult to reach. The result is that the natural and human resources of many poor countries have been exploited only to benefit the economies of the industrial and political powers of the world, such as the United States and the Soviet Union. Very often, the exploited countries do not even have the benefit of the short-term profits from the stripping of their land. Foreign companies and workers (aided by poorly paid and overworked native laborers) reap the bulk of the profits and transport the raw goods to their own homelands for further refinement. When they have finally used up the area's resources, they leave behind a polluted wasteland.

As always in history, international struggles continue, as nations attempt to control the wealth that lies beneath the earth. Nowhere is this more evident than in the frantic attempts to control the oil supplies of the Middle East and elsewhere. So intense is the buildup of emotions over the issue that it is commonly thought to be a prime potential cause of a major world conflict. Meanwhile, mine laborers continue to struggle for a fair wage and decent standard of living. Although great strides have been made in the West, the struggle for miners' rights in the East—notably in the communist bloc nations of Eastern Europe—lags far behind.

For related occupations in this volume, *Manufacturers and Miners*, see the following:
    Metalsmiths
    Power and Fuel Merchants
    Well Diggers and Drillers

For related occupations in other volumes of the series, see the following:
in *Artists and Artisans*:
    Potters
in *Builders*:
    Construction Laborers
    Masons
    Roadbuilders
in *Financiers and Traders*:
    Bankers and Financiers
    Merchants and Shopkeepers
in *Harvesters*:
    Farmers
in *Helpers and Aides*:
    Drivers
    Movers
    Private Guards and Detectives
in *Leaders and Lawyers*:
    Inspectors
    Political Leaders
in *Restaurateurs and Innkeepers*:
    Bakers and Millers
in *Scholars and Priests*:
    Monks and Nuns

in *Scientists and Technologists*:
   Engineers
   Geologists
in *Warriors and Adventurers*:
   Robbers and Other Criminals
   Soldiers

Power and Fuel Merchants

The people who have supplied fuel and power to energize mechanical devices and manufacturing processes have had a long history. Those who have done so for public use have had a significant role in the profession only in recent times. The ancient Egyptians and other early peoples used primarily human and animal power for public works and manufacturing. Many Roman slave owners rented out the services of their *laborers*, and may, in this respect, be thought of as professional suppliers of energy. The Chinese developed special collars and shoes for horses, which made the animals a reliable source of power. These innovations finally reached Europe by the ninth century A.D., and horsepower created a virtual revolution in farming, transportation, and industry, which was unrivaled until the advent of the steam engine in the 19th century A.D. Owners of horses rented and traded their animals for use in laboring and transportation.

The first large-scale harnessing of natural energy came in the form of the *waterwheel*, which dated back to ancient times, and the *windmill*, which came into general use in the Late Middle Ages. As early as the second century B.C., the Egyptians had made use of a *noria*, an automatic irrigation waterwheel. In the first century B.C. Vitruvius gave instructions for an innovative waterwheel that actually used gears to turn millstones at a much higher speed of rotation than the wheel itself. Similar devices were used as far north as Denmark and as far east as China by the following century. In these early times, however, people used little imagination in the use of waterpower. It was not until the 10th century, records seem to indicate, that mills were put to use in any way other than grinding grain into flour. By 1008, a document listing the donation of properties to a monastery in Milan notes mills used for both grinding grain and fulling (a process by which cloth is shrunk and thickened). In Germany, they were used for operating water-driven trip-hammers in mines and forges, and the *Domesday Book* notes some 5,624 mills operating in England in 1086. Sawmills, paint mills, and mills for tanning leather were soon using waterwheel power, and water-worked bellows for ironworks were used on the Continent and in England.

The windmill was the next great source of natural energy to be harnessed for man's use. Some say it was inspired by boat sails, such as those used in northwest Europe and Scandinavia in the early Middle Ages. Others suggest that the windmill was inspired by the Tibetan prayer wheels and Chinese revolving bookcases, which were used in Buddhist monasteries by at least the ninth century A.D. Whatever its source of inspiration, the windmill existed in Afghanistan by the 10th century

A.D., in Europe by at least the 12th century A.D., and in China by the 13th century A.D. The first windmills probably had actual sails, like those used on ships. By the early 14th century, Europeans had largely replaced waterpower with wind power, particularly in the northern plains. Windmills had revolutionized the textile industry by making hand or foot fulling outmoded (as waterwheels had already done in some areas). Wind power had come to be the main source of energy in mills used for tanning, laundering, sawing, hammering iron, and operating blast-furnace bellows; it was also used for grinding and polishing weapons, armor, and even diamonds and gold, as well as for mashing pigments into paint, pulp into paper, and grain into beer.

The merchants of waterpower and wind power were typically the owners, builders, and designers of the mills. A great many mills were rented out for a profit by landowners or merchants. When waterpower was used, *millowners* had to locate on rivers or broad streams, whereas windmills opened up the use of the plains. Sometimes *millwrights*—that is, *mill designers* and *engineers*—were sought to help set up the mill and its operations. They played a key role in the strategy employed for the harnessing of energy. Millwrights and mill designers were often wandering consultants, working for specified fees or favors. In many cases they were little more than *carpenters* and *masons*.

The millowners were the people who provided the capital and facilities for harnessing natural energy that could ultimately be converted into mechanical power. Although they were not solely in business to provide energy, as major power companies in the modern world are, they provided such power almost single-handedly. Of course, not all owners were individuals. Monasteries and cities both undertook the expense of constructing mills in order to finance manufacturing and processing ventures. In harnessing natural energy—either for themselves or for others who rented their mills—private, monastic, and municipal owners of mills set the basis for the Commercial Revolution. Out of this came a world that would eventually come to rely heavily on professional *fuel* and *power merchants*.

Water mills had lost much of their usefulness by the 16th century. Finding it difficult to always find suitable river locations for their operations, some waterwheel owners mounted their gear on boats or floating ramps, while others tried to harness tidal energy. These forms proved largely impractical, though, due to the periodic freezing of rivers and the congestion of boat traffic, and because of seasonal alterations of tide levels. Windmills remained common until the development of the *steam engine* in the 19th century. On the banks of the Zoon in the heart of Holland's famous industrial district, nearly 1,000 windmills were being operated by both millowners and renters by the end of the 17th century. At the turn of the 20th century, twice that number were in operation

Charcoal makers cut down wood and processed it in kilns to convert it into charcoal for fuel. (From Diderot's Encyclopedia, *late 18th century)*

throughout Holland. The steam engine and other more sophisticated mechanical devices eventually replaced the windmill, with the final blow being the harnessing of electrical energy.

Fuels had satisfied some of the energy needs of manufacturing for many centuries. The first large-scale use of fuels was in Britain in the late Middle Ages and into the 17th century. Professional *charcoal makers—lumberers* for the most part—provided wood-based fuel that, when burned, served *metalsmiths, forgers, glassblowers*, and *limeburners*. Charcoal was used where intense heat was needed more than simple mechanical energy. Charcoal makers prospered from wood byproducts as well. They sold bark to *tanners*, and alkali from wood ash for use in processes such as soap boiling, glassblowing, and cloth making. The Commercial Revolution created greater demands for wood, mostly in association with water transport, which was an essential ingredient in the new flurry of trade activity. *Shipbuilders* joined the charcoal makers in stripping the forests of England, where charcoal making was most fully developed as an occupation. Tar and pitch were extracted to better preserve the wood of oceangoing vessels, and willow charcoal was sold for gunpowder. The making of ships and guns had both greatly stimulated charcoal making, but only for a short time.

The wholesale depletion of large tracts of forest lands soon posed serious environmental problems in England, where most of this commercial and industrial activity was being conducted. Shipbuilders and charcoal makers began to fight for forest rights, but the royal government was much more interested in its naval than its industrial power at that time. In 1558 it became illegal for the iron industry to cut forests within 14 miles of any coastline. The new law was intended to conserve fertile coastal soil and prevent massive runoffs. By the end of the 16th century charcoal makers were well on their way out of business, and one contemporary observer, Edmund Howes, summarized the sorry situation:

> . . . within man's memory, it was held impossible to have any want of wood in England. But . . . such hath been the great expence of timber for navigation, with infinite increase of building of houses, with the great expence of wood to make household furniture, casks, and other vessels not to be numbered, and of carts, wagons and coaches, besides the extreme waste of wood in making iron, burning of brick and tiles, that at this present, through the great consuming of wood as aforesaid, and the neglect of planting of woods, there is so great a scarcity of wood throughout the whole kingdom that not only the City of London, all haven-towns and in very many parts within the land, the inhabitants in general are constrained to make their fires of sea-coal or pit-coal, even in the chambers of honourable personages, and through necessity which is the mother of all arts, they have late years devised the making of iron, the making of all sorts of glass and burning of bricks with sea-coal or pit-coal.

The story of charcoal making is a significant one because it summarizes the main issue posed by the fuel industry ever since: the desire to exploit natural resources versus the need to protect the environment. The Industrial Revolution was powered mostly by coal, the extraction of which proved to be far more economical and less damaging to the surface environment than the production of charcoal had been. Other problems soon faced the coal producers, though. By the middle of the 19th century there was growing concern for the health hazard of inhaling coal fumes and smoke. Oil became a welcomed substitute in the 20th century, because it was much cleaner to burn and easier to extract. In recent times, the shortage of oil in the world has created considerable international tension, as the industrial nations scramble to accumulate dwindling supplies. The Middle East, the greatest single source of the Earth's remaining oil, has become the center of this tension. Oil producers there have come to dominate the world's political scene since 1970 with their price setting and control of supply.

The entire situation has prompted many nations and power companies to seek alternative fuels and energy sources. As a result, coal producers are once again finding more receptive markets for their more economical fuel—although usually at the expense of clean-air standards. *Miners* and *drillers*, who find and supply energy in the forms of coal, oil, and natural gas, are dealt with in separate articles, as are *whalers*, who provided whale oil for lighting before the use of petroleum and electricity.

Public energy is a very new concept. Up until the late 19th century, it barely existed. Other than the energy used in manufacturing and transportation, little was generated at all. Domestic use in private households was negligible. Heating was accomplished by the use of stoves and fireplaces, which burned wood or coal; lighting was provided by torches, candles, or kerosene lamps. City streets were generally unlit, except for occasional torches placed in busy night-life districts. The well-to-do were once accompanied by private *torchbearers* while walking dark city streets. An ordinance passed in 1408 documents an early attempt to illuminate city streets in Paris; several others followed in later years. The following excerpt from one in 1541 was typical:

> And since the greater number of misdeeds are committed at night and in the darkness, all proprietors of houses shall every evening before six o'clock during the months of November, December and January place a lantern above the windows of the first story in a convenient and visible place, with a lighted candle to give light in the street, under penalty of 20 *sols parisis* fine

Most of these attempts were unenforceable and failed to make streets safer even where they were honored, due to the poor quality of lighting from candles and lanterns. An ordinance in 1588 ordered *falots* (vessels containing combustible materials like tar and resin) to be burned at all Paris street crossings between 6 P.M. and 4 A.M.

People were appointed to light, snuff, and otherwise maintain the *falots*; these may have been the first professional *lamplighters*. They were apparently not very successful in carrying out their duties, however, and Paris remained essentially unlit for several centuries to come.

London was one of the first cities in modern history to illuminate its streets. The 18th-century Paving Acts, well ahead of their time in attempts to tend to the sanitation, maintenance, and safety of British streets, provided for the appointment of *trustees* and *commissioners* empowered to light the streets under their care. This was quite an improvement over London's Act of Common Council of 1716, which commanded householders to hang out torchlights in the winter months between 6 and 11 P.M. on "dark nights." By the end of the 18th century, London had the best-lighted streets in the world, and regular *lamplighters* had been hired to tend to them. By the end of the 19th century nearly every parish had its own lamps and lamplighters. These parish lamps and their tenders were described in retrospect by one critic:

> Forty years ago the lighting of the streets was effected by what were called "parish lamps." The lamp consisted of a small tin vessel, half filled with the worst train oil, that the parochial authorities, for the most part the chosen of the select vestries, could purchase at the lowest price to themselves and the highest charge to the rate payers. In this fluid fish-blubber was a piece of cotton twist which formed the wick. A set of greasy fellows redolent of Greenland Dock were employed to trim and light these lamps, which they accomplished by the apparatus of a formidable pair of scissors, a flaming flambeau of pitched rope and a rickety ladder, to the annoyance and danger of all passersby. The oil vessel and wick were enclosed in a case of semi-opaque glass . . . which obscured even the little light it encircled.

Lamplighters were employed well into the 20th century in areas that electricity had not yet reached—rural districts and underdeveloped countries. By and large, though, electricity transmitted through power plants had preempted the lamplighting profession by the beginning of the century. Power for private consumers became the center of a viable enterprise shortly after Thomas Edison invented the electric light bulb in

The dirty, sweaty labor of many workers was needed to provide fuel for growing cities, as here at the Lambeth Gas Works. (By Gustave Doré, from London: A Pilgrimage, *1872)*

1879. The first actual power plant supplying electrical energy to consumers began operating in 1882 on Holborn Viaduct, London; the second began some months later in New York City. The electric utility industry subsequently grew until it served virtually every household, business, and industry in the technologically developed world. Many businesses such as railways, hospitals, and factories generate their own electricity, but the public electric utilities are still one of the largest industries in the world today. Most of them also now control the transmission of natural gas for heating. Many generate power through the controversial use of nuclear reactors, while most use traditional steam generators.

Public power utilities may be privately owned or government owned. The great majority of those in the United States are privately owned, but the governments of other countries often play a much more aggressive role in the industry. Power companies are major employers, and their rise has also led indirectly to growth in independent but closely related occupations.

The *electrician* prepares a household or business for receiving power by connecting wires and power boxes in the appropriate manner; the *plumber* sets up heating systems, which will eventually be activated by natural gas or electricity from a power plant, or from a home or plant boiler fueled and often maintained by a private oil company. The utility companies themselves hire thousands of people in many different capacities. *System planners* try to analyze trends in power usage as well as ecological, economic, and technical factors that may affect the long-term planning of the power system and its anticipated growth. *Operating engineers* are responsible for the daily function and maintenance of the company and its facilities. They must handle emergency situations such as power failures and be familiar with the amount of the plant's reserve generating capacity, and how to tap it, as well as system repair procedures. *Substation supervisors* handle the operation of intermediate relay stations.

A great many people deal more directly with the

delivery of electricity and gas and serving customers. *Powerhouse* and *substation electricians* and *mechanics* work on the wiring and machines of the main plant and its substations; *electricity* and *gas dispatchers* are responsible for the proper discharge of power to consumers; *customer-service representatives* deal directly with consumers regarding billing procedures, usage reports, and service complaints; *furnace* and *boiler installers* and *inspectors, meter installers* and *inspectors*, and *gas-leak inspectors* are responsible for the proper functioning of machinery directly delivering power on location to consumers; *meter readers* check periodically to see how much gas or electricity is being consumed for billing purposes; *cable installers* and *repairers, streetlight servicers, pole inspectors* and *line repairers*, often called *linemen*, work directly on cables, lamps, poles, and lines that carry power from stations to consumers; *leak-gangs* and their supervisors check for and repair gas leaks; *troubleshooters* attempt to locate sources of electrical power "breaks" and make the necessary repairs to restore service to customers; and *tree trimmers* help keep electric cables and wires free from obstruction or damage from protruding or weak trees, branches, or shrubs.

Power and energy have become extremely expensive in recent years, due largely to haphazard, careless, and wasteful use and the shortage of fuels. Many people are now attempting to avoid the high rates of public utility companies by generating their own power and by supplying their own fuel. Some now burn wood for home heating, while others have tried to trap wind or solar energy to activate heating, cooling, and generator systems. Many companies now sell a wide variety of items to help home or business owners in these ventures. Fireplaces, fuel savers, home windmills, and passive solar systems are devices leading the way in spurring new, independent business interest in the fuel and power enterprise.

For related occupations in this volume, *Manufacturers and Miners*, see the following:

Factory Workers
   Mechanics and Repairers
   Metalsmiths
   Weapon Makers
   Well Diggers and Drillers

For related occupations in other volumes of the series, see the following:
in *Artists and Artisans*:
   Glassblowers
   Jewelers
in *Builders*:
   Architects and Contractors
   Carpenters
   Construction Laborers
   Masons
   Plasterers and Other Finishing Workers
   Plumbers
   Roadbuilders
   Shipwrights
in *Clothiers*:
   Dyers and Other Cloth Finishers
   Fiber Workers
   Shoemakers and Other Leatherworkers
in *Communicators*:
   Printers
in *Financiers and Traders*:
   Merchants and Shopkeepers
in *Harvesters*:
   Farmers
   Whalers
in *Helpers and Aides*:
   Launderers
in *Restaurateurs and Innkeepers*:
   Bakers and Millers
   Brewers
in *Scientists and Technologists*:
   Engineers

Vehicle Makers

Vehicles were crude and rare in the ancient world. Wooden-wheeled carts and wagons were used in the Indus River Valley and in Central Asia by 2500 B.C., in Mesopotamia and Egypt by 2000 B.C., and in China and Scandinavia by at least 1300 B.C. After about 2000 B.C., chariots became fashionable with the Hittites—one of the earliest cultures to make actual military use of the horse. The Greeks and Romans used chariots, but not so much for military purposes as for racing. In most ancient societies, vehicle making was not a specialty craft, although country *wheelwrights* operated lucrative shops from early times. The earliest wheelwrights made crude, solid-wood wheels for farm carts, and frequently made the whole cart. In China, covered wagons were manufactured by teams of wheelwrights, *metalworkers*, and *carpenters*.

The vehicle-making profession grew somewhat during Greek and especially Roman times, as roads and postal services improved dramatically. Chariot makers were often civil servants or slaves attached to military units, although some were private artisans—such as wheelwrights or carpenters. The trade was neither prestigious nor particularly lucrative, except for the wheelwright. And wheelwrights' products were not made primarily for vehicles, but rather for gristmills and water pumps, as the Greeks and Romans began to make extensive use of geared wheels for private milling and public waterworks.

The first truly specialized craftspeople of the vehicle-making profession did not appear in society until modern times in Europe. After the Roman Empire declined, most of the great roadways fell into disrepair. Overland hauling reverted to the use of pack animals, and *mail carriers* and travelers had to go on horseback or on foot. In the late Middle Ages, wagons and carts began to be used more frequently as trade increased, but it was the fine carriage trade begun in the late 16th and early 17th centuries that really made vehicle construction a

Wagonwrights made carts, wheels, wagons, and carriages, as well as wooden implements like plows. (By Jost Amman, from The Book of Trades, *late 16th century)*

significant craft and commercial enterprise for the first time.

Coaches and carriages differed significantly from wagons and carts in both construction and purpose. The latter were built in three parts—the body, forecarriage, and hindcarriage—by country wheelwrights and carpenters. They were bulky, heavy, and uncomfortable, and made primarily for hauling farm produce and other goods into marketplaces. Coaches and carriages, first built in the 16th century on the Continent in Hungary and elsewhere, were built in only two parts—the body and a one-piece carriage—by town coach builders. Suspended by leather straps for cushioning and shock absorption, and built lighter and higher off the ground than farm vehicles, carriages were used for the leisure travel of aristocrats and the high-speed exploits of royal mail carriers. As great cities sprang up all over Europe, travel became leisurely and vehicle ownership prestigious for the first time in history. Master *joiners* or *coachbuilders* came to practice a rather leisurely urban craft in the service of an elite clientele, and they were paid well for the contracts they filled. They had to hire many assistants and craftworkers in related specialties: wheelwrights, *blacksmiths*, ornamental *ironworkers, carvers, upholsterers*, and *leatherworkers*. The management of this team effort eventually came to demand so much time and energy that the well-known coach builders themselves rarely did any work on the carriages that bore their names.

Walter Rippon built the first British carriages in 1555: one for the Earl of Rutland and a second one for Mary Tudor. During the Elizabethan era, carriage travel became exceedingly popular among British nobility, and many cheaper sedan chairs and hackney coaches were manufactured for the less-affluent folk and for public use. The poet John Taylor complained that the carriages and sedans being built at that time "tost, tumbled, rumbled and jumbled without mercy," while endangering the *boatman*'s livelihood, though Samuel Pepys noted that the Thames River was still England's chief highway three-quarters of the way through the 17th century. Carriage making, in any case, was becoming a thriving industry made up of many crafts. Rarely did a single worker build an entire coach from start to finish. In fact, the undertaking was one of the earliest examples of specialization and the importance of coordinated teamwork before the Industrial Age, while still deeply rooted in the era of handcrafting.

When considering the creation of the London Company of Coachmakers and Coach Harness Makers, it was originally thought that both the wheelwrights and coach makers (of which the *coach harness makers* were a part) should form a single guild. The idea was finally approved on the condition that the company "shall not molest or compel any person whatsoever free of any Company within this City using the trade of a wheelwright or coachmaker to be translated in their corporation." In time, of course, it was hardly possible to practice the trade at all unless one were a member of the company. But that was not necessarily bad for either the public or the profession, considering that the guild was created largely to overcome earlier common "fraudes and deceipts" of the unregulated "wheelwrights and Coachmakers." A committee formed by the Court of Aldermen at Guildhall reported in 1632 that it was those very unregulated craftsmen:

> . . . which doe use in their workes much yonge & unseasoned Tymber, which when it is fitted for wheeles or framed for coaches doe shrink and thereby being disioynted [disjointed] both Coach and wheeles quickly decaye, whereby men's occasions in their iornys [journeys] are disapoynted, & their lives many tymes endaungered. Wee finde alsoe that much deade wood in joynt worke of Coaches is used, which breaketh with the shaking of the Coach, and that these abuses, by reason that the said Coaches and Wheeles are paynted and collered, cannot easily be discovered but by one of those

facultys nor be reformed but by some politick constitucon prohibiting the same.

Carriage bodies were suspended by steel springs shortly before the American War of Independence. Rising standards of living in France, England, and even America were making the coach makers and their associates truly an elite class. It had been written as early as 1677 in the original Charter of the Company of Coachmakers and Coach Harness Makers, that the "Arts, trades and misteries of coach making are . . . of great use and beneficial to the noblemen of Our Kingdom and Other subjects." Certainly this held true in Europe in the 17th century, and in America by the late 18th century. Still, there were many other vehicle makers of less repute earning a living in the trade. Many *innkeepers* who operated post and coach stations got involved in the business of hauling goods and storing and renting vehicles for public and private use. Some became adept at repair work, and others began to actually construct wagons and stagecoaches.

In America, Boston was the center of the gentry's custom trade in vehicles. A native of the town—Major Adino Paddock—was the most reputable *chaise maker* in the Colonies before the American Revolution. The aristocracy was not so prominent in America, though, and the trade lagged throughout the Colonial era, with European-born artisans running most of the profitable shops. Coach makers had to be somewhat innovative to survive under such circumstances. A Jonathan Brown of Hartford got the edge on his chief rival, Consider Burt, by purchasing decorative coach trimmings from Boston rather than hiring *artists* to do that time-consuming part of the production. Even Paddock had to resort to selling second-hand chaises "under their value," while also offering to "Take Old Chaises in part pay for New." He also constructed sleighs and carriage parts for competitors and smaller dealers like Brown of Hartford.

Making carriages and coaches demanded a substantial capital outlay, so that only fairly well-to-do joiners (fine workers in wood) could get such a business off the ground anyway. Innkeepers would sometimes take the profits from their main business and invest them in a carriage shop. Most carriage builders were hard-pressed to cut overhead and labor costs. Late 18th-century British masters were charged by one observer as being "not content with a moderate profit upon a part they are unable to execute themselves." This reference was made to the pitifully low salaries paid to body painters and *limners*, who painted the family symbols, called the heraldry, on the coach. These were usually down-and-out *artists* unable to get any other work. Their labors in furnishing "oil, gold and colours" greatly increased the price that carriage makers could demand of their wealthy customers, yet none of that profit was passed down to the laboring artist. On the other hand, consumers also complained about the high cost of the carriage. One Englishman even worried that joiners were engaged in an "unreasonable abuse of the Nobility and Gentry of the Nation," though it is hard to sympathize too much with aristocrats who would not manage to live within their means.

The 19th century brought about vast changes in the industry. Thomas Telford and John Macadam had led the way to vast improvements in roadway systems and construction. Steel springs in carriages made riding more comfortable. Lighter carriages, such as the Italian *curricle*, the construction of which was originally patronized by the Prince of Wales, made carriage making more practical, simple, and lucrative. The guild system gradually disappeared, as the craft became more of an industrial enterprise. The following account by a German visitor to the premises of a British coach maker and harness maker illustrates the workings of the industry in the late 18th century. This particular establishment was half a century ahead of its time, with conditions more typical of the 19th-century carriage factory:

We went to Hatchett, one of London's most famous master saddlers, who employs several hundred workmen in his service. At home we have no conception of such a saddler, with premises for cartwrights, smiths, harness-makers, carvers, painters, upholsterers, gilders—all kinds of workmen necessary for coach and harness-making, and other accessories, working under his supervision and producing the loveliest masterpieces of their kind. I cannot think of any visit more interesting than this one: think of three floors of spacious rooms filled with swarms of busy people, whose perfect workmanship is only excelled by still more perfect implements.

The painters and lacquer-workers were on the third floor—all the main flights of stairs are broad, and so arranged that the banisters may be taken down, and the finished vehicles allowed to slide down in ropes. I especially admired the neat craftsmanship of the harness-workers and upholsterers. We concluded our tour amongst a number of finished coaches, and with an inspection of some fine drawings of all kinds of vehicles.

I was amused to see how the people played into each other's hands, as the saying goes; and that a saddler has a counting house and a paymaster just like a banker.

Improvements in mechanization and technical knowledge, along with better roadways, turned the 19th century into a golden era for carriage makers. In America, the horse and buggy became a national emblem. The British *gig*, the French *chaise*, and the Italian *curricle* all had their heydays on the production line, as more of the rising middle class placed orders for them. The new trend toward urban living—with its improved standard of living, better road systems, and leisure time for country rides—made the carriage business so prosperous that firms began construction even before custom orders were placed. The standardization of parts and even whole vehicles sped up the process, getting more carriages out to the public, and more money into the hands of carriage builders and their employees.

The downfall of the carriage trade was right around

the corner, however. International vehicle exhibitions were beginning to feature horseless carriages—automobiles. In 1896, a leader in the carriage-making industry warned those in his profession that "we should remain neutral, and be in a position to supply carriages for any motor that is introduced or required by a customer." Somewhat later *The Financial News* of London insisted that "carriage builders will have to convert themselves partially into engineers." For a long time, though, carriage builders fought the automobile and its makers, until it became undeniably clear that the days of the horse and carriage were past. Only then did many of the carriage builders try to get work in the coachworks of automobile companies. Most coach makers simply went out of business, and many actually sold their working premises to the automobile manufacturers—the historic Holland & Holland in England being a prime example.

The automobile industry and the aircraft industry brought about vast changes within the vehicle construction business. None was more revolutionary than Henry Ford's introduction of *assembly-line production* in the United States before World War I. From that day to this, the automobile and aircraft industries have paved the way for increasingly specific part assembly and specialization of tasks.

Automobile designers and aircraft designers are *engineers* rather than artisans. The industries also require many other categories of skilled, semi-skilled, and unskilled workers to produce a finished product. While the carriage maker of old was involved in all or most stages of both design and production, modern aircraft makers and automobile makers typically deal with nothing beyond their small area of expert knowledge. Indeed, in recent decades, they have become so far removed from the final production of a whole piece, and the target of overly aggressive "speedups"—that is, pressure for more and more rapid production. Some *social psychologists* have blamed the process for causing

Modern vehicles, like the fighter planes these noses are intended for, are generally produced in large factories. (National Archives, Records of the Office of War Information, 208-AA-352QQ-5, Douglas Aircraft, California, c. 1943)

worker apathy and nervousness, leading to poor quality of work to antisocial and even criminal behavior.

Automobile production skyrocketed between the two world wars, as did the number of companies engaged in the business. Before long, though, only a handful of huge trusts came to monopolize the industry. Automobile manufacturing lent considerable support to a number of other enterprises: the making of steel, rubber, and machine tools; the construction of roads, bridges, and tunnels; and the extraction and processing of petroleum. Attempts to organize labor in the industry, then, quickly had national and even international repercussions. The automobile workers made their first serious attempt to unionize in 1926, but were turned back by Ford and the other huge conglomerates. The United Automobile Workers (UAW) was finally founded in 1935, and after a series of bloody strikes in Flint, Michigan, won the recognition of General Motors. Walter Reuther, John L. Lewis, and others led the workers to further victories, until even Ford permitted them into the shop in 1941. Ever since, they have continued to gain personal dignity, job security, and some control over the running of the business, in the face of rapidly developing automation.

Today, automobile workers and aircraft workers continue to be paced by assembly-line production schemes in plants around the world. Nonetheless, workers cannot keep up with improvements in automation, even—or especially—in Japan, which is today a leader in automobile construction. One-armed robots can weld panels at least 50 times faster than even the best worker. Meanwhile, the high cost of both production and fuel, plus the shaky state of the world's economy in recent years, have caused a considerable slowdown in the manufacture of aircraft and motor vehicles. Labor unions within these professions have sometimes added to this sorry situation by insisting on ever-higher wages and greater benefits for their workers. The results have been a large number of plant shutdowns and worker layoffs. In some cases, workers have bought out the premises and machinery of failing factories and run the plants themselves to avert a complete shutdown. The workers who have retained their jobs are far removed from the craft techniques of the carriage builders. *Engineers* and *designers*, on the other hand, rarely have anything to do with the actual manufacture of the vehicles. Many of these jobs tend to offer less human satisfaction than the jobs of vehicle makers in the 18th and 19th century did. Only a few craftworkers engaged in the restoration or replication of old cars, carriages, and airplanes still enjoy the art of vehicle making as it once existed.

For related occupations in this volume, *Manufacturers and Miners*, see the following:
   Factory Workers
   Metalsmiths

For related occupations in other volumes of the series, see the following:
in *Artists and Artisans*:
   Furniture Makers

Painters
　　Sculptors
in *Builders*:
　　Carpenters
in *Clothiers*:
　　Shoemakers and Other Leatherworkers
　　Tailors and Dressmakers
in *Communicators*:
　　Messengers and Couriers
in *Helpers and Aides*:
　　Drivers
　　Movers
in *Restaurateurs and Innkeepers*:
　　Bakers and Millers
　　Innkeepers
in *Scientists and Technologists*:
　　Engineers

Weapon Makers

From the earliest times, the design and construction of the implements of war were undertaken by a variety of *artisans, chemists, scientists, theoreticians*, and *manufacturers*. Probably the first of these craftspeople were the prehistoric *stoneworkers* and *flintworkers*. In many societies, stoneworking was a skill everyone learned. But—some being better than others—specialties developed and expert stoneworkers would trade their weapons for other items. Working in stone was an extremely difficult, time-consuming chore. Stone weapons may be, in modern terms, primitive, but they are far from crude. Obsidian, flint, and some other types of stone hold a very fine edge. Indeed, some modern surgical instruments are made from these materials, since volcanic glass maintains a finer edge than even carbon steel. The disadvantage, of course, is that stone, once broken, cannot be mended.

With the discovery of metal, stone was replaced by the more workable material. Metal not only could be mended, but also could be used to make longer, more dangerous blades. The use of metal involved a huge technological step, for mining and smelting had to be mastered first. Copper was the first metal to be used for weapons in Egypt and India, where there are deposits of almost-pure copper. Copper was easy to work with, but was also very soft. So the early *sword makers* turned to creating alloys by mixing different metals. The first major alloy used was bronze, made from copper and tin. Later came iron and then steel, which was harder than copper and could be cast into molds. Iron ore was much more common than other metals used, making iron weapons the least expensive.

The sword was the paramount weapon in most early cultures, and a great deal of mystery and lore was attached to their manufacture. The most skilled *blacksmiths* devoted their lives to developing strong, flexible blades that would stay sharp. The Norse

The making of a sword—here a short-bladed assegai *being forged by a Zulu blacksmith—involved considerable skill. (From* Peoples of the World, *19th century)*

worshipped their swords, giving them names, since they were considered to be living entities. The most prized swords were believed to be made by dwarves in hidden caves. Dwarves were supposed to be the finest craftspeople because it was assumed that their size gave them unmatched dexterity.

Norse swords were painstakingly made by pattern welding and were exceptionally strong, well balanced, and durable. But perhaps the finest swords were made by the Japanese. Their smiths developed a method by which they would fold and refold the steel upon itself. Their blades were wonderfully balanced and so sharp that they could easily slice through a man. Swords hundreds of years old appear to be almost new.

Another group of particularly skilled weapon makers were those who created missile weapons, especially the *boyers* who made the bows and the *fletchers* who fashioned the arrows and strings. A skilled boyer would select a staff, study it, season it for over a year, and then shape it along the lines of the wood. A fine bow was a prized weapon, and only the best *archers* were even allowed to shoot the preferred and most expensive yew bows (particularly those made from Spanish yew). Meanwhile fletchers developed the arrows and strings to complete the ensemble, although *archers* could construct their own ammunition in a pinch. Of the European countries, boyers and fletchers were most important in England, where the free yeomanry (independent small farmers) provided a large source of trained archers. Some Scottish and English kings even banned football in order to try to force their subjects to practice archery. The French also were successful in introducing the bow as an effective weapon for their armies.

Weapon makers around the world poured a great deal of skill into bows. In the Middle East and Asia, bow making reached a very high stage of art. Craftspeople there used wood, horn, and sinew to shape the so-called composite bow, which demanded a high degree of technical expertise to create. The Turks were among the best at

Vulcan, the god of fire, was often pictured making arms and armor for the gods. (From Museum of Antiquity, *by L.W. Yaggy and T.L. Haines, 1882)*

making these weapons; their composite bows were capable of firing light arrows more than 800 yards.

Technical progress continued in the ancient art of weapon making, even when it had all but ceased in other enterprises—testimony to the importance of war to most early people. The legendary armorers, Vulcan and Wayland the Smith, were given high place in mythology. Few ancient arms makers made major careers out of their crafts, however. The famous Greek dramatist, Sophocles, was the son of Sophillus, an Athenian armorer who grew rich operating a thriving workshop, but he was certainly an exception to the rule. Most makers of weaponry and armor in antiquity also manufactured other metal wares and woodenwares, such as kitchen utensils, wagon wheels, and the like. Still, few earned very considerable fortunes for their efforts; most could barely stay in business at all. Even though war was practically a way of life for early people, weaponry was simple and armory only crudely developed. Aside from the sword makers, few artisans in the profession attained any notable repute. Indeed, for centuries, many *warriors* fashioned their own weapons, and it was usually only royal armies that could afford to retain special artisans to arm, protect, and equip their troops.

It was during the Middle Ages that weapon makers and armorers first developed into an elite class of artisans. The European institution of martial contests and tournaments during the much-heralded Age of Chivalry made the armorer a highly respectable craftworker and artist. At about the same time, the archers and boyers of England formed rather elite craft guilds. Sword makers and *cutlers* (those who made knives) remained in high esteem. They, too, had craft guilds in many a medieval European town, and remained the chief representatives of the profession in the West.

Personal, hand-to-hand combat was the chief method of fighting throughout the medieval world, and remained so well into the 16th century. It was the armorer who held the top seat among the weaponsmiths during this age, even though armor was only defensive gear, and played no role in actually inflicting harm on the enemy. Royal armorers were extremely important and extravagantly patronized by lords, kings, and nobles, whose presence and safety during the progress of a battle was of supreme significance. Armorers were well rewarded not only for the strength and utility of their products, but just as much for the decorative design that would distinguish the king and his army from the enemy. Even the royal horses were outfitted with beautifully worked coats of armor, which were works of art in their own right.

For many centuries armorers formed a small, elite group of wealthy craftworkers. It was not an easy group to join. Common metalworkers were excluded by a variety of contrived rules and requirements. In order to apply for master status, for example, a German apprentice in the craft had to present and turn over to the guild a personally owned and manufactured suit of armor. Ostensibly, the purpose of this stipulation was for the guild to be able to carefully inspect the quality of his work; in actuality, it eliminated a great many hopefuls from ever applying. A suit of armor was an expensive item, and only those of comfortable means could own one, much less surrender it just on the hope that they might gain admission into such prestigious company.

While armorers may have been the most prestigious weapon makers before modern times, the works of many other specialists within the profession were eagerly sought and dearly paid for. In Europe, certain localities became noted as manufacturing centers for particular weapons. This fame usually reflected either the fine reputation of a resident master *smith*, or the local availability of a given metal or other natural resource. The sword makers of Toulouse and Bordeaux in France, Solingen in Germany, Toledo in Spain, and Passau in Austria were considered among the finest in the world, as were the Versey makers of *misericordes* (daggers used to "mercifully" finish off a wounded soldier), the Catheloigne makers of *arbalests* (powerful crossbows), and the Colin crafters of *cleavers* (long-bladed hatchets). Even special pieces of armor were often produced separately by various specialists. Craftworkers in Milan were considered the finest makers of *hauberks* (shoulder-and-neck mails or even full-coat mails that were worn under the outer coat of heavy armor); and the *bucklers* of Barcelona were among the greatest shield makers and helmet makers.

Not all weaponsmiths were specialists in these early times; this was true of only the finest masters. Most of the more common craftworkers in the profession produced a wide variety of weapons. The same workshop might turn out crude armor, swords, helmets, and shields—even experimental firearms, after the Chinese invention of gunpowder in the 10th century spread to the West in the 14th century. It was only where particular crafts were regulated and protected by very strong guilds, such as those in England for fletchers and boyers, that real specialties typically developed.

The finest weapon makers of the Western world were generally from Germany, where raw metals and natural resources were in plentiful supply, and from Italy, where many of the secrets of Eastern and Islamic techniques and workmanship were introduced during the Renaissance.

Most weapon makers were primarily metalsmiths. They had to learn the art of tempering metals above all else. It is from this task that the thousands of jealously guarded craft secrets were developed. Some masters claimed to use miraculous springs for cooling, or poisoned ones to fashion especially lethal swords, or divinely graced suits of armor to protect important lords and kings in battle. Because of these mysterious secrets of tempering, forging, and plating, arms makers were regarded with some sense of awe, particularly considering the high stakes of war and hand-to-hand combat for national and personal survival. Yet, makers of weapons were basically not very different from any other of the metalsmiths. They busied themselves in hot workshops tending to furnaces, beating out plates from solid ingots of metal, and finally shaping the finished product into suitable armor or weapons, generally into an artistic piece that any warrior would be proud to display.

Those who traded and financed the manufacture of arms, rather than the actual forgers and artisans producing them, made the greatest profits from war, of course. *Miners, forgers*, and *metal dealers* provided master armorers and weaponsmiths with raw materials and metal ingots from which rough—only sometimes finished—products were fashioned. The weapons and armor were then returned to the mines and forges, where they were often finished and then prepared for marketing. The largest profits went into the hands of the actual *traders* of the finished weapons. Many of these traders were mercilessly unscrupulous, and most were more interested in building their private fortunes than in supporting local or national causes.

The modern age of weapon making did not begin until reliable cannon and firearms could be produced. For several centuries after their invention, guns and cannon were of less importance in warfare than were such weapons as lances, armor, swords, and bows and arrows. During this time, *gunsmiths* were not as specialized, organized, or prestigious as armorers, boyers, fletchers, and sword makers. Guns and cannon had been made by various metalworkers and blacksmiths, as well as by *cannon and gun founders*, at least as early as the 14th century B.C. But the expense, danger, imperfections, and inefficiency of these weapons rendered them of little general use for centuries. As a result, the Chinese invention of gunpowder had few real effects on the arms industry until the modern Age of Discovery, when Europeans began moving into the unknown world. Then guns, crude as they were, allowed Europeans to conquer with relative ease great empires, whose people had no such weapons.

The modern age of weapon making also was made possible by mechanization, the standardization of parts, mass production, and finally, industrialization. Guns helped to speed this process along by inspiring—and benefiting from—many industrial and technological innovations. Mechanization in armor production began at a surprisingly early time. Even before the middle of the 15th century, Milanese armorers were operating factories that, in just a few days, could arm an entire infantry of 2,000 men and a cavalry of 4,000 men and horses. Armorers of this and later periods employed water-powered battering mills to mass-produce rough armor, mail, and weapons. Within one workshop, *hammermen, millmen, platers, mailers*, and *armorers* all represented distinct divisions of labor and often operated different water-powered tools, machines, and other devices. The even more complex nature of firearms and cannon called for even greater mechanization and standardization within the profession of arms manufacturing.

The first gun and cannon producers were armorers, founders, and forgers who had the technical knowledge of metalworking and the machinery and implements for banging, plating, melting, and casting rough ores into huge cannon and standard firearms. As early as the 14th century, gunpowder was being used in Europe to fire projectiles in metal casings. Military engineers, who had previously given most of their attention to the building of catapults and battering rams to knock down enemy defenses, joined with *metallurgists* and *alchemists* in exploring the use of gunpowder toward the same end but with greater efficiency and effectiveness.

Cannon were extremely powerful weapons, whose effectiveness depended to some extent on what they were loaded with. Soldiers aiming to blow a hole in a wall or knock down a gun turret would probably use one massive cannon ball. But if they were shooting at an army coming over a hill, they would be more likely to load the cannon with many tiny balls, called *shot*, or with jagged pieces of metal, called *shrapnel*. Weapon makers made something of a science of developing items that, when shot from a cannon, would do the most damage. Some, for example, developed linked pieces of metal designed to cut through and knock down the masts of a ship.

For some time, the outdated efforts of the small-scale general weaponsmith competed successfully with those of the more modern gun and cannon manufacturers. One reason was that craft guilds retained a considerable portion of their medieval influence within the industry and over local politics. The guilds tried to forbid foreign competition, and to some extent succeeded, in many cases right up to the 17th century. Regardless, royal and noble armies continued to demand the services of German and Italian armorers. Kings themselves continued to patronize master weapon makers and armorers, enjoying the personal service they gave.

But the major reason that the arms industry remained archaic for so long was the general mistrust of gunpowder and firearms. Military engineers dabbling in chemistry and metallurgy experimented wildly and often with disastrous results. They made gunpowder from willow charcoal, saltpeter, and sulphur, never quite

The sword maker's shop was frequented long after the advent of firearms, for gentlemen continued to carry swords as signs of their station. (From Diderot's Encyclopedia, *late 18th century)*

sure of the results of their experiments. Early *gunfounders* were thought of as eccentrics and dangerous alchemists whose works were "weak, broken, noisome, used up and broken and wasted in trials and assays," according to one observer in the middle of the 14th century. Guns and cannon long remained small, dangerous, and unpredictable. Muskets were cumbersome and had to be supported by forked rods while a *fuse firer* discharged them. How to light the charge in small arms was a question that completely confounded arms makers for many years. Progress in perfecting lock, trigger, and firing breaches was painfully slow, as the matchlock was replaced by the wheel lock, which in turn was replaced by the flintlock in the days of George Washington's military service. Even the flintlock, though, often misfired and was useless in the rain. Many of the users of firearms were killed by misfires, and the weapons' unpredictablity posed a constant danger to anyone near them. Moreover, hand-to-hand combat was considered more courageous and honorable by both European aristocrats and Eastern swordsmen. The Japanese "Bushido" (Way of the Warrior) doctrine, for example, held that only face-to-face combat with steel weapons was befitting a true warrior. The use of gunpowder was thought to be cowardly and no indication of a man's strength or fighting ability.

While gunsmiths and cannon founders were having trouble marketing their wares, other weapon makers and armorers were little better off. Despite the existence of protective guilds, the armaments industry was a highly competitive one. The most prosperous people in the profession were those who won royal patronage and were thus given the task of outfitting and arming the king and the royal troops. Masters were often, however, in the unenviable position of having their good and knowledgeable advice disregarded, while still being held accountable for the final results. Holy Roman Emperor Maximilian once scolded his master Conrad Seusenhofer by saying: "Arm me according to my own wish, for it is I and not you who will take part in the tournament."

While a royal post held a high degree of prestige, it did not always pay well. Arms makers were rather like *painters* and *sculptors*, who labored endlessly for royal and noble patrons, but in the end had to beg and plead for the wages promised to them. Countless private armsmakers and arms dealers went bankrupt for failure to collect on government contracts that they had too eagerly filled. Eventually, they became somewhat wiser, and a great many insisted on cash on delivery for arms ordered during wartime. Good but cold business practices like these rapidly earned arms makers and arms dealers a reputation for being miserly, opportunistic, and even unpatriotic.

Guns and cannon manufactured by arms makers in Liege, in Belgium, supplied the entire army of Charles the Bold of Burgundy, who used them during his attack against Belgium and in his destruction of Liege itself. Obviously, this harmed the civic reputation of the industrious and affluent arms manufacturers of that key city. Nonetheless, the international traffic in arms—condemned by many popular observers as antinational—continued. In 1576, during the Roman Catholic Counter-Reformation, the Duke of Alva and his Spanish troops flooded the Low Countries and massacred a great number of Protestants with guns and cannon that had been manufactured in Liege. The Germans tried to stop the arms makers of Liege from shipping weapons to their enemies, as did Napoleon and the French some years later. But the armament manufacturers of that city continued to flourish and prosper without any real interruption, all the while selling to whichever country paid the best price on the open market.

This sort of situation was very common in the arms industry. Queen Elizabeth of England granted special privileges and monopolies to two arms manufacturers in Sussex, yet by the end of the 16th century they were directing a massive black-market trade with foreign countries. In all, four of every five cannons manufactured in Britain were finding their way overseas. In the following century, Swedish founders joined the British in secretly supplying arms to various European nations. Arms manufacturers were getting rich on the perpetuation of war everywhere, and they hardly seemed to care who was fighting, why they were fighting, or who might win. European governments, appalled by this callous capitalistic and opportunistic attitude, seemed helpless in their attempts to influence the giants of the industry.

In early munitions factories, machines like these were developed to pierce the barrels of large guns. (From Diderot's Encyclopedia, *late 18th century)*

Many governments decided to take control of the manufacture of armaments. Under Louis XIV of France, Colbert began to license armament firms while creating state manufactories. Most of the muskets and gunpowder made in France during the 17th and 18th centuries were produced in state factories by workers who were in reality civil servants. A Prussian musket factory employing some 250 men was founded in 1722 at Spandau. Russian arms factories at Sestroreck and Toula were worked by about 700 men each. In the United States, the Virginia Armory took over government contracts for guns and weapons at the end of the 18th century.

Like most state arms manufactories, the Virginia Armory sought to lure skilled gun masters and weaponsmiths from private practices and workshops and into civil service. The greatest incentive offered was that private weapon makers could no longer hope to obtain government contracts. The Virginia Armory was run by a superintendent, a master armorer, an assistant master armorer, and two clerks above the regular work force. George Williamson, appointed the armory's first master armorer in 1802, was granted an annual salary of $1,000. He had been a private arms manufacturer ever since the American Revolution and maintained a gun shop in Henrico County, Virginia, where he had also served as justice of the peace since 1799. His background was typical of such offices, and he squarely fit the bill that superintendent Clark had issued concerning the search for master armorers: They "should be active, impartial men, well skilled in the art of making of their own hands all the several parts of the arms to be manufactured."

Of course, the greatest problem for state arms factories was in attracting talented artificers who might be able to do ironwork on the production machinery as well as turn out quality arms and parts. Slave labor was apparently not used for this purpose, and local gunsmiths—already in short supply to fill private and hunting contracts—earned more money in independent rifle shops. Governor James Monroe corresponded with Clark over this issue, and he decided that it was best to try to entice Irish immigrants from New England and Pennsylvania, even though he recognized that "the very circumstance of bringing them from a distance . . . is calculated to inspire them with too high an idea of their merit, and will incur some expence not applicable to those in the neighborhood." The workers finally obtained for the Virginia Armory received a typical contract in May 1803. According to its terms, they were employed for a trial period of one month, after which they might be retained for a period of three months to three years. They were obliged to spend some time guarding the powder house, had to wear an approved uniform or apron, and could be discharged at any time for improper conduct or poor quality of work.

While some master armorers and gunsmiths went to work at state factories like the Virginia Armory, the future for the industry lay primarily in the development of private factories. Even in the 18th century, great independent manufacturers dominated the industry in Germany. Georg von Giesche cast cannon in brass, Jean Martin Wendel mastered the use of iron and steel in producing armor, and the Stumm family turned out the highest-quality armor plating in the world. By the end of the century the British had taken over the traditional continental role as the greatest gunmakers. This was the finest period in the art of British gunmaking. The London Gunmaker's Company, a strict and closed union, boasted a membership that included the best master gunsmiths in the world.

A typical successful British gunmaker of this period was John Twigg. He spent seven years as an official apprentice and another five years as a journeyman before becoming an independent gunmaker in 1755. He became rich, not by seeking or filling contracts for the gentry and nobility—most of whom rarely paid their bills and caused many a fine armorer's bankruptcy—but from contracts for muskets and pistols from the East India Company. Twigg passed on his coveted trade secrets and great fortune to his own family (particularly his nephew, John Bass), which dominated the industry for many more years.

The art of weapon making and gunsmithing was rapidly fading, though, and master craftworkers increasingly became—or were replaced by—factory and assembly-line owners and operators rather than artists and artificers. By the time the War of 1812 was being fought, Eli Whitney had introduced the mass production of muskets, and not long after that the first modern revolver was invented by Samuel Colt. The Colt revolver was soon manufactured on an assembly line, which was made all the more efficient by the production of interchangeable parts, standardized for the first time. The repeating Winchester rifle entered the scene in the 1860's, along with Remington's rapid-load carbines. All of these technical innovations put weapon makers at the forefront of the Industrial Age. The whole focus of war changed as weapons became more deadly and more efficient over unbelievably longer distances, thus "depersonalizing" the act of killing.

The weapon-makers who ran the private factories that manufactured guns and bombs as well as swords and knives began to reap fortunes that had never before been known in the industry. Early in the 19th century, the *shipbuilder* was profiting the most from war. But the day of the arms manufacturer was dawning. Most of the "new breed" in the industry were not master artisans at all. Rather, they were businessmen. Most of them had operated businesses that had little or nothing to do with arms before the revolution in armaments manufacture began. In England, the Vickers works—in the 20th century one of the greatest arms makers in the world—was founded in 1828, but began making guns out of steel only in 1869. Other great names in the industry traced similar paths. The Armstrong factory founded in England in 1846 produced only hydroelectric machinery until the Crimean War inspired its operators to begin making guns; the German Krupp foundry (established in 1812) manufactured only steel utensils until guns and rifles came to dominate its operations 30 years later; the French Schneiders made only steam engines for 20 years before their production plans turned completely to armament; and the Skoda works of Bohemia (founded in 1866) was simply a machine factory for 20 years and only attached an ammunition factory in 1913.

Many of the leaders of the armament industries were politically active. Eugene Schneider, for instance, was the Grand Officier of the French Legion of Honor. He used his position to win large arms contracts in 1853 during the Crimean War. Karl Skoda was made an honorary baron, and he also used his influence to obtain large contracts. In fact, most of the great arms producers of the late 19th and early 20th century seemed to be working as hard within political circles to get wars started or to keep them going as they were in the actual production of armaments. The wealthy old man Krupp, scolded by the Kaiser for supplying both Austria and Prussia during the Austro-Prussian War, retorted:

> We [the Krupp armament factories] cannot live on Prussia alone; for the next ten years we shall need at least 50 millions' worth of orders. And if foreign countries place orders with me, I cannot decently send them goods of inferior quality.

The "antinational" international trade in arms was out of control. If Krupp could defend his seemingly unpatriotic actions as simply "decent" and good business, who else could be blamed for similar practices? During the Franco-Prussian War, the American government itself helped the Remington Company of Connecticut get orders from the French. The Colt plant in London made massive shipments to Russia, supplying both sides in the Crimean War. Remington also supplied both sides of the Russo-Turkish War in 1879, as the following observation makes clear:

Then Russia and Turkey decided to fight. Both patronized the Bridgeport factory, and the strange situation developed of one plant daily grinding out thousands of cartridges for the combatants to fire against each other in deadly battle. Both nations had their inspectors at the works. The officers treated each other with formal courtesy while they inspected millions of the little messengers of death which were to fill the air of Southeastern Europe with noise and destruction.

This trend continued in the 20th century. Vickers of England became the main supplier of the Japanese Steel Works. Between the two world wars, several arms manufacturers banded together to undermine the attempt of Western nations to negotiate a general disarmament and international peace. Bethlehem Steel and other steel conglomerates sent William B. Shearer to Geneva in 1927 to denounce the disarmaments talks, but he became so abusive in his zeal to thwart peace that he had to be recalled. The same arms manufacturers also hit hard on the home front, encouraging "hawkish" propaganda through politicians and the press. In all these instances, the special-interest spokespeople of the weapon makers played on the spirit of nationalism and the fear of other countries to push programs of full militarism.

Many major arms merchants continued to use their influence shrewdly to help bolster business. In the 1930's, M. Schneider, president of a major arms producer—Creusot—held virtual control over the *Comité des Forges* (the French steel trust), which in turn maintained a controlling interest in the two greatest French newspapers: *Le Temps* and the *Journal des Débats*. It is no wonder that those papers preached nothing more vehemently than the dangers of disarmament and the threat of neighboring Germany. Also in the 1930's, several leading engineers of Vickers were tried in court for their part in the sale of 60 of the most modern and deadly British tanks to the Russians. The "War Scare of 1930" was perhaps the most shocking incident of the period, which shows just how far some arms merchants might go to help business. Zelevski, a Skoda agent in Romania, bribed officials of that country to project a false public fear of an imminent Russian invasion. After Skoda secured a large weapons order from Romania, the war scare completely dissipated.

In the latter part of the 19th century, another trend began in the arms industry. In order to become somewhat less dependent on constant warfare to meet profit projections, some arms manufacturers began to diversify their operations. Remington began selling typewriters, sewing machines, and farm machinery. Du Pont manufactured hunting equipment and blasting powder, and became active in clearing land for railroad building in the second half of the 19th century. Still, arms manufacturers relied heavily on war for profits, and they did all they could to monopolize that part of their business. In 1897 a joint American-European ammunition cartel was formed to standardize export prices and slow down cutthroat competition among themselves. In 1907, Du Pont was found to be in violation of the Sherman Anti-Trust Law. But during World War II, nearly half of all the U.S. government's orders for powder went to the Du Pont "Powder Trust."

There is no denying the enormous impact that weapon makers have had on history. In the United States alone, the settlement of the entire West depended on the existence of portable arms. These had little use in Europe, but were essential ingredients to the opening of the American frontier and the defeat of Mexican and Native Americans. While cannon founders were the most important manufacturers of arms in 19th-century Europe, the small arms and ammunition industry in America rapidly established itself as one of the first great independent arms industries in the world. From the time that Thomas Jefferson approved the establishment of the Du Pont plant in Delaware, to the Civil War, nearly 40 small arms and 5 ammunition manufacturers opened shop in the New World. This became the nucleus of a

During the two world wars, women moved into heavy factory work in arsenals, making the weapons necessary to fight the war. (National Archives, Office of the Chief Signal Officer, 111-SC-35757, 1918)

thriving industry that would soon see Colt, Winchester, and Remington arms in every corner of the world, dominating the international arms-traffic scene.

Since World War II and the development of nuclear weapons, portable arms and small artillery have become less important for national defense, although they remain extremely popular for private and recreational use. The arms merchants of the atomic age have increasingly become the operators of large conglomerates, chemical and steel corporations, and government defense departments themselves in many cases. Craft in making weapons has been replaced by the precision and complex calculation of teams of scientists, including *computer scientists* and engineers, who work on constantly improving and updating weaponry.

While the traffic in arms has become decidedly more nationalistic in recent years, there is still a flourishing trade among countries that is the seed of many a growing dispute on the international scene. Whether legal or not, moral or otherwise, arms manufacturers and arms dealers still seem prone to seek the best profits attainable for whatever national or social cause or use demand their products.

For related occupations in this volume, *Manufacturers and Miners*, see the following:
    Factory Workers
    Metalsmiths
    Miners and Quarriers

For related occupations in other volumes of the series, see the following:
in *Builders*:
    Architects and Contractors
    Shipwrights
in *Financiers and Traders*:
    Merchants and Shopkeepers
in *Scientists and Technologists*:
    Alchemists
    Chemists
    Computer Scientists
    Engineers
    Physicists
in *Warriors and Adventurers*:
    Sailors
    Soldiers

Well Diggers
and Drillers

Well diggers and *Drillers* have historically provided the public with water supplies for drinking, power, and transportation. Most recently, drilling for subsurface petroleum and natural gas has become a new and even more significant aspect of this profession.

In ancient times, *slaves* were used to work on massive public projects that called for large-scale digging operations. Wells were dug for water collection, conveyance, and storage. This was especially important in desert civilizations such as Egypt and Mesopotamia, which had precious little water to begin with and had to be very careful how they used and stored what they did have. As early as 2500 B.C., advanced water systems existed at Mohenjodaro and Harappa in the Indus River

Basin, including brick-lined wells and intricate drainage systems. The diggers were most likely slaves and *civil servants*. The Chinese had wells as deep as 1,500 feet, and the Greeks sank clay pipes in their tunnels to provide water supplies to the cities. The Romans were famous for their engineering feats, which included digging or building aqueducts. By 305 A.D. they had over 350 miles of water-carrying aqueducts and stone arches. The Romans had public baths and rest rooms and so many wells and conduits that they needed a full-time *water commissioner*.

Most ancient and medieval wells were little more than shallow cavities dug in the vicinity of a well spring. But in 1582 a power pump was first used to go deep into the floor of the Thames River at the London Bridge. From that time, digging became more and more mechanized. Slaves had not been used for digging for many centuries, even though there were extensive waterworks with aqueducts and lead pipes in Paris, London, and Hanover. Free labor was used, and most men were paid pitifully low wages for their long hours of backbreaking work. Debtors and common *criminals* were frequently used to supply digging labor. They were rented by private companies or set to work on public service projects. *Dowsers* were widely employed, usually on a free-lance basis, to find the proper spot to begin digging for water. They used a *divining rod* to detect the source of underground springs. The crooked, forked stick supposedly bent toward water sources, but most people credit the dowser's success to superior knowledge and instinct.

In 1699 a private digging company in London began running underground pipes to private houses. Later in that century, cast-iron pipes began to be used, improving and therefore stimulating the well-digging business. Great demands for wells and water systems were forthcoming. Labor shortages forced wages higher, but the digger was still grossly exploited and ill-treated. Boston established the first American city waterworks in

1652, although private wells had been dug from the time that the first settlers arrived at Jamestown. There were public wells in New York City during the Revolutionary era, but the quality of the water was so bad that *street vendors* made a booming business of selling "tea-water" for drinking—which was nothing more than upstate spring water.

In 1761 steam pumps were installed in London's waterworks system. They made water much more accessible to the general public and further stimulated large-scale well-digging ventures. Steam pumps were used in Paris by 1781 and in Philadelphia by 1800, when only 16 city water systems existed in the entire United States.

The 19th century saw a much greater demand for public and private well systems. Urban areas were growing rapidly with the Industrial Revolution which, at the same time, brought new mechanization and refined techniques for the digging and drilling of wells. By the beginning of the 20th century nearly every town in the United States had a public water system. Today wells are opened with power drills, making for easier but more calculating work. The well digger has largely been replaced by the highly trained driller, and the business of well digging and drilling is much less labor-intensive than ever before.

The greatest source of employment for drillers today is in the petroleum and natural gas industries. Early in the 19th century an oil extract from rocks was used as a sort of kerosene to fuel lamps. The first deliberately dug commercial oil well in the world was drilled in 1859 by Colonel Edwin Drake at Oil Creek in Titusville, Pennsylvania. Oil was so valuable as a burning fuel in lamps (electricity had not yet been harnessed for domestic or public use) that an oil rush ensued. *Wildcatters*—get-rich-quick oil prospectors—flooded Pennsylvania in 1859 in search of more gushers. Most were disappointed but many were not. By the end of the century John D. Rockefeller had established one of the world's greatest single producers of petroleum—the Standard Oil Company.

Many *oilers* have become rich since the 20th century, as oil has become one of the most valuable resources on Earth. Transportation, home heating, and commercial heating have all become overwhelmingly dependent on the fuel. By 1969 there were 600,000 producing oil wells in over 60 countries—principally in the Middle East and the Soviet Union. Natural gas, which often was extracted with oil, was originally burned on site, because there were relatively few uses for it and it was treated as a by-product. In the last 50 years, however, natural gas has become nearly as important a fuel as petroleum, and many drilling companies spend millions of dollars annually in searching for and extracting it.

Many people are employed throughout the world in the business of drilling for oil and natural gas. The job has become particularly sophisticated in recent years, demanding the services of highly trained *drilling, production*, and *reservoir engineers* as well as machine and computer operators. Since 71 percent of the Earth's surface is covered by water, a significant number of drilling operations are probing the ocean floor for oil, natural gas, and even manganese, which is used in steel production. The Frasch pumping process has made deep-sea explorations more practical. Workers using this and similar techniques live on huge steel platforms, which tower over 60 feet above water and are connected by bridges. They get to their stations by helicopter. Once there, they remain for months, without returning home to their families or private lives. They are housed comfortably in air-conditioned units, but have dangerous and lonely jobs. There is always the risk of a platform disaster in which support columns malfunction, costing the lives of many crew members. The compensation for such dangerous and lonely work is high pay, which enables workers to live uncommonly sumptuous life-styles when they are home.

Many of the jobs related to drilling operations are less glamorous. They are the common laboring jobs of chipping paint off tankers and loading filthy barges with heavy drill pipe. No skill is needed, and in some countries such work is done by drifters and *day laborers*. In the Oil Patch of the United States, which stretches from Louisiana to Texas, bunkhouses were built during the boom years of the industry; they are still used today to house wandering laborers looking for an honest day's work. These men often arrive penniless and are given food and shelter at a cost that is deducted from their future paychecks. Many bunkhouses are little more than breeding grounds for crime and violence or hideaways for fugitives. Many represent the grossest form of the exploitation of common labor. Yet others provide a valuable service in supplying companies with needed labor and giving otherwise homeless and unemployed vagabonds a place to sleep, eat, and work—at least for awhile.

Oil and natural gas extraction involves many types of workers, from common laborers to highly trained

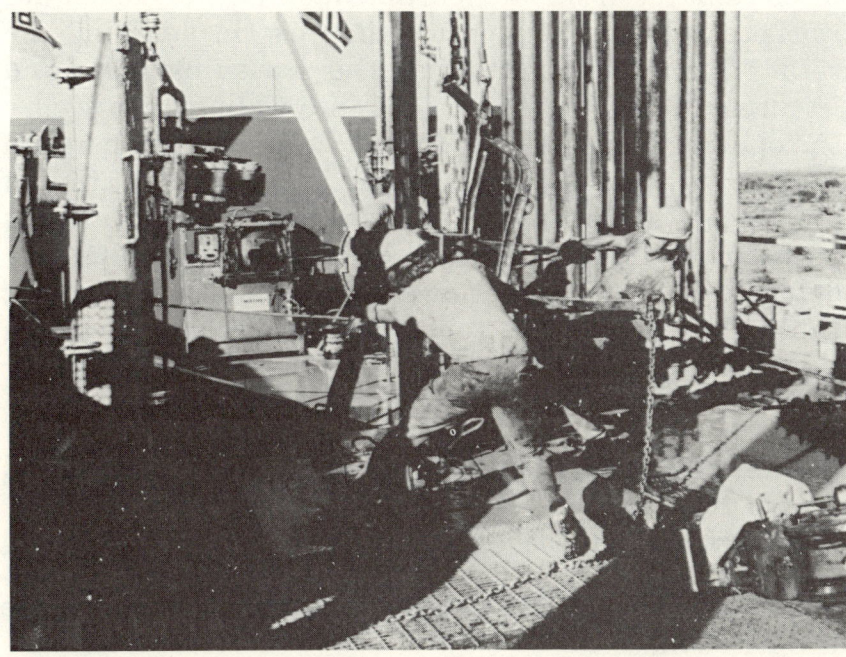

Oil workers on the floor of a drilling southeast of Phoenix, Arizona. (Phillips Petroleum Company, from American Petroleum Insitute)

professionals. *Geologists* and *geophysicists* are employed in the exploration for oil and gas. They map subsurface rock formations and suggest appropriate areas to begin drilling operations. Teams of drillers are headed by specialist engineers, and pipes are laid by laborers working under the direction of *civil engineers*. Countless other professions are involved in the refining, processing, and transport of oil.

Laborers are classified into many categories. *Prospecting drillers* drill shallow boreholes to obtain geological and crust formation samples to help in assessing the feasibility of working a certain site. *Roughnecks* are responsible for the assembly and repair of machinery and drilling equipment; they also dig holes, connect frameworks, and dismantle and assemble boilers and steam engines. *Roustabouts* do odd labor, usually with hand tools; they help dig building and drainage trenches, clear surface areas to be drilled, mix concrete, and load and unload trucks. *Derrick workers* operate power hoists to transfer rod sections (used for well casings) to and from derrick racks, as the rods are placed in or removed from wells. *Catheads* and *lead-tongs* are workers named for the tools they use in assisting the rotary driller by tightening and loosening bolts and fixtures. *Casers* "stab" sections into wellholes to secure the opening to be worked. *Swabbers* and *bailers* perform various functions in the cleaning out and restoration of old and overworked oil and gas wells. *Line walkers* patrol pipelines by foot, horse, camel, or automobile; they detect, report, and sometimes repair leaks or breakages; they also inspect and repair telephone and telegraph lines used by the drilling company. *Blasters* provide, place, and detonate explosives to help open up wells.

For related occupations in this volume, *Manufacturers and Miners*, see the following:

Miners and Quarriers

For related occupations in other volumes of the series, see the following:

in *Builders*:
    Construction Laborers
    Plumbers
in *Restaurateurs and Innkeepers*:
    Costermongers and Grocers
in *Scientists and Technologists*:
    Computer Scientists
    Engineers
    Geologists
    Physicists
in *Warriors and Adventurers*:
    Robbers and Criminals

Suggestions for Further Reading

For further information about the occupations in this family, you may wish to consult the books below.

General

The American Worker (The United States Department of Labor Bicentennial History of . . .). Edited by Richard B. Morris. Washington, D.C.: U.S. Government Printing Office, 1976. A fine pictorial journey through the history of American labor, its work conditions, and organized movements.

Brooks, Thomas R. *Toil and Trouble: A History of American Labor*, 2nd ed. New York: Dell, 1971. A general review of the progress made by American workers in several industries.

Cahn, William. *A Pictorial History of American Labor*. New York: Crown, 1972. A colorful account of the struggles of labor in American history.

Gimpel, Jean. *The Medieval Machine: The Industrial Revolution of the Middle Ages*. New York: Holt, Rinehart and Winston, 1976. A general account of medieval industry.

Warner, George Townsend. *Landmarks in English Industrial History*. New York: Macmillan, 1898. A detailed review of the evolving role of the industrial worker in England.

Factory Workers

Dodd, George. *Days at the Factories; or the Manufacturing Industry of Great Britain Described, and Illustrated by Numerous Engravings of Machines and Processes*. East Ardsley, England: EP Publishing Limited, 1975; reprint of 1843 edition. A detailed look at manufacturing processes and labor organizations in Regency era England.

Huggett, Frank E. *A Day in The Life of a Victorian Factory Worker*. London: George Allen & Unwin, 1973. A re-creation intended primarily for students.

Kydd, Samuel. *The History of the Factory Movement*. New York: August M. Kelley, Publishers, 1966; reprint of 1857 edition of Simpkin, Marshall, London. An economic classic describing in detail, and through original accounts and testimonies, the plight of factory workers in

Great Britain from 1802 until the enactment of the Ten Hours' Bill in 1847.

Metalsmiths and Miners

Agricola, Georgius. *De Re Metallica*. New York: Dover, 1950. Translated from the first Latin edition of 1556 by Herbert Clark Hoover and Lou Henry Hoover. The classic work on the medieval metallurgical industry; detailed and intriguing.

Knauth, Percy. *The Metalsmiths*. New York: Time-Life Books, 1974. An excellent survey of the art and technique of the ancient metalworkers throughout the world; includes good illustrations and photographs of actual artifacts and work methods.

Roadbuilders and Engineers

Hindley, Geoffrey. *A History of Roads*. Secaucus, New Jersey: Citadel Press, 1972. A thorough and readable treatment of the subject, with considerable material on roadbuilders, designers, and engineers, covering prehistoric to modern times.

Weapon Makers

Canby, Courtlandt. *A History of Weaponry*. New York: Hawthorne, 1963. An illustrated look at the evolution of war-making tools.

Cromwell, Giles. *The Virginia Manufactory of Arms*. Charlottesville: University Press of Virginia, 1975. A complete history of the Virginia Armory, and the professional manufacture of arms as a transition from crafts to standardized mass production.

Dupuy, T.N. *The Evolution of Weapons and Warfare*. Indianapolis: Bobbs-Merrill, 1980. An intelligent discussion of the development of warfare.

Engelbrecht, H.C., and F.C. Hanighen. *Merchants of Death: A Study of the International Armament Industry*. New York: Dodd, Mead, 1934. An extremely readable and vivid exposé of the extent of the operations and influence of the arms makers, emphasizing the great European manufacturers of the 19th and early 20th centuries.

Ffoulkes, Charles. *The Armourer and His Craft: From the XI to the XVI Century*. New York: Benjamin Blom, 1912. A detailed account of the medieval and early modern armorers and their materials and techniques; includes helpful diagrams and plates.

Howe, Russell Warren. *Weapons: The International Game of Arms, Money and Diplomacy*. Garden City, New York: Doubleday, 1980. Includes a short historical piece on weapon manufacturing, but emphasizes the contemporary industry and the moral implications of disarmament.

Jackson, Melvin H., and Charles de Beer. *Eighteenth Century Gunfounding*. Washington, D.C.: Smithsonian Institution Press, 1974. A detailed description of the art of gun founding as it first underwent technological change; emphasizes the Verbruggens in Holland and England.

Lewinsohn, Richard. *The Profits of War Through the Ages*. New York: Dutton, 1937. Tells of how people have made fortunes through war; nearly one-third of the book focuses on armament firms and also covers financiers of army supplies.

Neal, Keith W., and D.H.L. Back. *Great British Gunmakers: 1740-1890*. London: Sotheby Parke Bernet

Publications, 1975. A detailed account of some of the world's most outstanding 18th-century family armories and gunmakers, paying particular attention to John Twigg and the gun room of Packington Hall.

Oakeshott, R.E. *The Archaeology of Weapons*. New York: Praeger, 1960. A good description of the development of killing tools.

Payne-Gallway, Ralph. *The Crossbow: Mediaeval and Modern Military and Sporting: Its Construction, History and Management*. New York: Branhall House, 1963. An excellent, detailed work about the most-feared medieval weapon.

Robinson, H. Russell. *Oriental Armour*. New York: Walker, 1967. A fine work, concentrating on Asian armor.

Stone, George Cameron. *A Glossary of the Construction, Decoration and Use of Arms and Armor in All Countries and in All Times*. New York: Jack Brussel Publisher, 1934. A fine source book that describes virtually every imaginable weapon; particularly good for Asian weapons.

Index

Aerospace mechanics, 33
African miners, 92
Age of Chivalry, 129
Age of Discovery, 132
Agricola, Georgius, 1, 80
Aircraft designers and industry, 120
Alchemists, 1-2, 133
Alva, Duke of, 135-136
Amon, 64
Appalachian coal mining, 96
Arbalests, 130
Archaeologists, 35
Archers, 127
Armorers, 129-130, 132
Arms makers, 135-136
Armstrong factory, 139
Artists, 118
Ashley (Lord), 90
Assayers and assaying, 1-3

Assembly-line production, 12
Australian gold rush, 92
Austrian copper mines, 68
Automated plants, 25
Automobile workers and automobiles, 32, 33, 120
    unionization, 121-122

Bailers, 150
Bankers, 1
Bass, John, 139
Bauxite, 97
Belgian arms makers, 135-136
Bell-founders, 53
Bellows, 41
Berkman, Alexander, 22
Bessemer converter, 49
Bessemer, Henry, 89
Bethlehem Steel, 141
Biringuiccio, Vannoccio, 1, 47-48

Black labor, 93
Blacksmiths, 2, 9, 30, 38, 44, 46, 55-56, 58, 115, 126
Blast furnaces, 44
Blasters, 150
Bloomeries, 37, 41, 44
Blowers, 42
Blowpipes, 41
Blue-collar workers, 5
Boiler installers and inspectors, 111
Bootmakers, 20
Boyers, 127-128
Brassworkers: see Braziers
Brazers, 59
Braziers, 53-54
Bridge-stocker, 51
British Isles: see England
Bronze Age in Egypt, 63
Bronzeworkers and bronze, 36, 40, 44
Brown, Jonathan, 117
Brown, Willie, 91
Bucklers, 130
Burt, Consider, 117
Bushido doctrine, 134
Byzantine factory workers, 8

Cabinetmaker, 30
Cable installers and repairers, 111
Cannon and gun founders, 132, 133
Capitalists, 47
Carnegie, Andrew, 51
Carpenters, 9, 30, 40, 103, 113, 115
Carriages: see Coaches and carriages
Carthusian monks, 43
Carvers, 115
Casers, 150
Casting carriers, 58
Catheads, 150
Caucasians, 66
Cave of the Treasure, 66
Certification mechanics, 34
Chaise makers, 117, 118
Charcoal makers, 48-49, 105-106
Chariot makers, 114
Charles the Bold, 135
Chemists, 41-42
Children and child labor laws, 18, 20
China
  coal, 67, 81
  gunpowder, 130, 132
  metalsmiths, 39, 40
  power merchants, 101
  steel, 38
  vehicle makers, 113
Chokedamp, 82
Cistercian monks, 8-9, 43, 75-76

Civil engineers, 149
Civil servants, 146
Claireveux, Abbey of, 43, 76
Cleavers, 130
Clockmaking, 11
Closed shops, 95
Coach harness makers, 116
Coachbuilders, 115
Coaches and carriages, 114-120
  carriage trade, 114-115
  horseless carriages: see Automobile workers and automobiles
Coal miners and coal mining, 48-49, 81-86
  dangers, 82-86
  labor unions, 88
  United States, 88-89
  women, 84-85
Cofferers, 11
Collective bargaining, 23
Colonial America
  metalsmiths, 49-50
  miners, 86-87
Colt revolver, 139
Colt, Samuel, 139
Comite des Forges, 141
Commercial Revolution
  factory workers, 12
  metalsmiths, 44
  power and fuel merchants, 105
Commissioners, 108
Common Council, Act of, 108
Communist countries, factory workers in, 24
Company of Coachmakers and Coach Harness Makers, 116, 117
Company of Undertakers for the Iron Works, 87
Computer industry mechanics and scientists, 33, 144
Conversi, 43
Conveyors, 59
Convict labor, 93
Cooks, 29
Copper and copper mining, 36, 39, 63, 65-66, 68, 87, 126
Coppersmiths, 52, 53-54
Corbie, Abbey of, 8-9
Cordwainers, 11
Coremakers, 58
Craftspeople, 6, 11
Crane, Stephen, 94-95
Criminals, 146
Crucible, 2
Cupel, 2
Curricle, 118, 118
Curriers, 11

Customer service representatives, 111
Cutlers, 129

Damascus swords, 37, 38
Dangers
  coal miners, 82-86
  metalsmiths, 47-48
Davy's lamp, 91
Day laborers, 149
De Re Metallica, 80
Deep-sea mining, 97
Derrick workers, 150
Designers, 122
Diamonds, 67
Diderot, 13
Diggers, 72
Dirty work, 79
Divining rod, 146
Domesday Book, 102
Domestic servants, 20
Domestic system, 12-13
Dowsers, 146
Drake, Edwin (Colonel), 147
Drillers, 97, 107
Drop-hammer operators, 58
Du Pont, 142
Dwarfs, 42

E. Morgan's Sons soap manufactory, 15
Earthenware, 8
Edinburgh Review, 50-51
Edison, Thomas, 109-110
Egypt
  Bronze Age, 63
  copper processing, 65-66
  metalsmiths, 37-38, 39
  mining industry, 61-63
  quarrying, 62
  waterwheels, 102
Electricians and electricity, 109-111
Electricity dispatchers, 111
Electronics industry mechanics, 33
Elizabeth (queen of England), 136
Encyclopedia, 13
Engineers, 30, 70, 120, 122
  see also specific types of engineer, i.e., Civil engineers
England, 40
  archery, 127
  automobiles, 120-122
  carriage travel, 115
  coal miners, 90-91
  Commercial Revolution, 105-106

Common Council, Act of, 108
Factory Acts, 21
Fielden's Act, 90-91
flint mines, 61
guilds, 11, 116, 129
gunmaking, 138
Hobhouse's Act, 90
Industrial Revolution, 82-86
iron mining, 74
metalsmiths, 46-47, 50-51
Paving Acts, 108
steelworkers, 51
vehicle makers, 118-119
water systems, 146, 147
Ethiopia
  gold, 40
  mining industry, 64
European mechanics and repairers, 29-30
Exchangers, 1
Exploration engineers, 98
Extruder operators, 58

Factory Acts, 21
Factory workers, 5-26
Faggoting, 41
Falots, 107-108
Farmers, 7, 68
Farriers, 44
Fielden's Act, 90-91
Fillers, 42, 51
Financial News, The, 120
Finishers, 59
Firedamp, 83-84, 91
Fletchers, 127-128
Flintworkers and flint, 61, 125-126
Flowmeter testers, 34
Flynn, John, 94
Ford, Henry, 120, 121
Forgers and forging, 37, 38, 39, 58, 105, 131
Forgiae errantes, 45
Forging press operators, 58
Founders, 38, 39, 58
France, 73
  Comite des Forges, 141
  falots, 107-108
  St. Leu d'Esserent mines, 74
Franklin, Benjamin, 88
Frasch pumping process, 148
Fuel merchants, 101-111
  Commercial Revolution, 105
  Industrial Revolution, 106
Fuggers, 47
Furnaces
  installers and inspectors, 111
  managers, 51
  operators, 58

Fuse firer, 134

Gallagher, Patrick B., 94
Gas dispatchers, 111
Gas-leak inspectors, 111
General Motors, 121
Geological engineers, 98
Geologists, 150
Geophysicists, 149
Germany
  Federsee Lake community, 66
  Krupp foundry, 139-140
  miners and quarriers, 72-73, 75
  weapon makers, 138, 139-140
Gesner, Abraham (Dr.), 89
Giesche, Georg von, 138
Gig, 118
Girdlers, 11
Glassblowers, 105
Glovers, 11
Gold, 36, 37, 40, 61-62, 64, 71, 92
Gold rush, 87-88
Golden Fleece, 69
Goldsmiths, 1, 9, 42, 44
Gowen, Franklin B., 93
Grape harvesters, off-season, 7
Great Britain: *see* England
Greece
  factory workers, 6
  miners, 69
  vehicle makers, 114
Grinders, 30, 31
Guards, 68
Guilds, 12
  British, 129
  metalsmiths, 44, 53
  shopkeepers and traders, 31
  urban, 11-12
  vehicle makers, 116
Gunfounders, 134
Gunpowder, 81, 130, 132, 133-134
Gunsmiths, 132
Gynaecea, 8
Gypsy tinkers, 39

Hammermen, 132
Hammersmiths, 58
Handthrowsters, 12
Harrow, Isaac, 49-50
Hathor, 66
Hattusilis III (king of the Hittite), 65
Hauberks, 130
Herodotus, 40
Hittites
  iron production, 37-38, 64-65
  metalsmiths, 40
  mining industry, 61
Hobhouse's Act, 90

Holland & Holland, 120
Holland, steam engines in, 104-105
Horseless carriages: *see* Automobile workers and automobiles
Horsepower, 101-102
Horseshoer, 44
Hor-ur-Re, 63-64
Howes, Edmund, 105-106
Hungary
  metalsmiths, 39-40, 47
  vehicle makers, 115

Illnesses
  coal miners, 82-86
  factory workers, 20
India
  mining, 67
  steel, use of, 38
Industrial Revolution, 10, 106
  factory workers, 18-19
  mechanics and repairers, 31-32
  metalsmiths, 48-49
  miners and quarriers, 80-92
Industrial Workers of the World (IWW), 22
Ingots, 45-46
Innkeepers, 117, 118
Iran
  ancient mining community, 67-68
  copper, 39
Ireland; Coote's ironworks, 80-81
Iron, 36, 37, 40, 43, 44-45, 50-51, 74 seric, 38
Iron Age, 63
Iron man pick, 91
Ironmasters, 49
Ironmongers, 50
Ironsmiths, 37
Ironworkers, 115
Israeli metalsmiths, 38
Italian factory workers, 12

Jade, 67
Japan
  automated plants, 25
  automobile construction, 122
  Bushido doctrine, 134
  metalsmiths, 40
  swords, 126
Japanese Steel Works, 141
Jefferson, Thomas, 142
Jewelers, 42
Joiners, 115
Journal des Debats, 141

Keepers, 51
Kerosene, 89

Krupp foundry, 139-140

Labor unions
  automobile workers, 121-122
  coal miners, 88
Lace, Brussels, 14
Laird, Addie, 17
Lamplighters, 108-109
Lay-out inspectors, 58
Lead, 36
Lead-tongs, 150
Leak gangs, 111
Leather dyers, 11
Leatherworkers, 11, 115
Legislators, 23
Lewis, John L., 93-94, 95-96, 121-122
Limeburners, 105
Limners, 118
Line walkers, 150
Linemen, 111
Linen, 8
Loaders, 59
Locksmiths, 9
London Gunmaker's Company, 138
Longfellow, Henry Wadsworth, 55
Lorimers, 11
Louis XIV (king of France), 14, 137
Love, William, 94
Lumberers, 105

Macadam, John, 118
Machine molders, 58
Machinists, 59
Mail carriers, 114
Mailers, 132
Malemakers, 11
Manufactories, 6
Manufacture Royale at the Gobelins, 14
Markers, 59
Masons, 9, 103
Mass production; metalworking, 57-58
Match manufacturers, 20
Maximilian (Holy Roman emperor), 135
Mears (Messr.), 53
Mechanics, 29-34
  automobile, 32, 33
Merchants, 29
Metal dealers, 131
Metallurgical engineers, 98
Metallurgists, 133
Metalsmiths and metalworking, 30, 35-59, 73, 105, 113, 131
  Colonial America, 49-50
  Commercial Revolution, 44
  dangers, 47-48
  guilds, 44
  mass-production, 57-58
  Middle Ages, 42-43
  modern division of labor, 50-51
  noise created by, 46-47
  seasonal employment, 45
  traveling, 42
Meteorites, 38, 67
Meter installers and inspectors, 111
Meter readers, 111
Methane explosions: *see* Firedamp
Middle Ages
  metalsmiths, 42-43
  miners and quarriers, 71-80
  quarriers, 73-74
  weapon makers and armorers, 129-130
Military engineers, 133-134
Military mechanics, 33
Mill designers: *see* Millwrights
Mill operators, 12
Mill rollers, 58
Millers and mills, 29, 102
Millmen, 132
Millowners, 103
Millwrights, 103
Mine city, 77-78
Miners, 61-99, 107, 131
  Industrial Revolution, 80-92
  Middle Ages, 71-80
  modern times, 92-99
  see also specific types of mining, i.e., Coal mining; Strip mining
Mining administration, 77
Misericordes, 130
Modern times
  automobile construction by robot, 122
  factory workers, 25-26
  mechanics and repairers, 32
  metalworkers, 56
  miners and quarriers, 92-99
  steel industry, 57
Molders, 58
Moldmakers, 23
Molly Maguire organization, 93, 94
Monasteries, 10, 43
Monetary systems, 76
Money changers, 1
Monroe, James, 137-138
Muffle, 2
Mycenaeans, 69

Natural gas, 97-98, 148
  drillers, 147
Needlemakers, 13
Noria, 102
Norse swords, 126-127
North American colonies
  mechanics and repairers, 30-31
  mining, 86-87

Oil, 106-107
Oil mining, 97-98
Oil Patch, 149
Oilers, 148
Old Testament, Fifth Book of Deuteronomy, 65
Oliver, Andrew, 49
O'Neill, Thomas, 94
Open shops, 95
Operating engineers, 109

Paddock, Adino (major), 117
Parchment makers, 9
Patternmakers, 59
Paving Acts, 108
Peddlers, wandering, 30-31
Pepys, Samuel, 115-116
Petroleum industry, 147
Pewterers, 30, 46
Pig iron, 44, 49, 52
Pilers, 59
Pinkerton's Detective Agency, 94
Pirotechnia, 47-48
Platers, 132
Pliny the Elder, 68-69
Plumbers, 109
Poland; Solidarity movement, 24
Pole inspectors, 111
Porcelain, 8, 14
Porters, 42, 68
Potteries, urban, 10
Potters and pottery, 20, 37
Pouchmakers, 11
Power looms, 14
Power merchants, 101-111
  Commercial Revolution, 105
  Industrial Revolution, 106
Powerhouse electricians and mechanics, 111
Preparers, 68
Press maintainers, 33
Press setters, 59
Printing, 10
Production engineers, 98, 148
Production managers, 11
Prospecting drillers, 150
Prospectors, 2, 72
Putting-out system, 12-13

Quarriers, 61-99
  Industrial Revolution, 80-92
  Middle Ages, 71-80
  modern times, 92-99

Rackers, 59
Reaumur, Rene Antoine, 49
Remington Company, 140-141, 142
rapid-load carbines, 139
Renaissance
  European mechanics and repairers, 29-30
  metalsmiths, 44
Repairers, 29-34
Reservoir engineers, 148
Reuther, Walter, 121-122
Rippon, Walter, 115
Riveters, 59
Robots used in automobile construction, 122
Rockefeller, John D., 148
Roman Empire
  factory workers, 7-8
  miners and quarriers, 69-70
  vehicle makers, 114
Roughers, 59
Roughnecks, 150
Roustabouts, 150
Rubies, 67
Rutland, Earl of, 115

Saddlers, 11
Salt traders and mines, 71, 73
Scale makers, 46
Schneider, Eugene, 140, 141
Scoria, 2
Scrap balers, 59
Semi-factory organizations, 13
Seric iron, 38
Seusehofer, Conrad, 135
Shearer, William B., 141
Shipbuilders, 9, 105, 139
Shoemakers, 9, 29
Shopkeepers' guilds, 31
Short-timers, 18
Shot, 133
Shrapnel, 133
Silicosis, 86
Silk, 8
Silversmiths and silver, 36, 37, 44
Sinai Peninsula; mining industry, 62, 63
Skinners, 11
Skoda, Karl, 140, 142
Slaves, 6, 29, 37, 41, 42, 63, 87, 89, 145, 146
Smelters and smelting, 37, 68

Smith, Adam, 13-14
Smiths, 30, 41
Social psychologists, 120-121
Solderers, 59
Soldiers, 37, 63, 79
Solidarity movement, 24
Sophillus, 128
Sophocles, 128
South American metalsmiths, 40
Spanish miners, 70
Spectrographic analysis, 3
Standard Oil Company, 148
Stanneries, 78
Steam engines, 104-105
Steam pumps, 147
Steelmakers and steel, 36-37,
    50, 51, 57, 59
  Bessemer, 49
  wootz, 38
  see also Pig iron
Stewards, 9
Stocktakers, 51
Stoneworkers and stone, 73-74,
    125-126
Stow, John, 46
Strabo, 68-69
Street vendors, 147
Streetlight servicers, 111
Strip mining, 96, 97
Stumm family, 138
Substations
  electricians and mechanics, 111
  supervisors, 109
Sudanese gold mines, 71
Suetonius, 7-8
Sumerian mining, 67
Surface mining, 97
Survey of London, 46
Swabbers, 150
Sword makers, 126-127
System planners, 109

Tacitus, 72
Tanners, 11, 105
Taylor, John, 115
Technicians, 33
Television repairers, 33
Telford, Thomas, 118
Temps, Le, 141
Tenters, 18
Textile factories, 10-11, 14
  women, 17-18
Timberers, 66
Tinkers, 31, 39
Tinsmiths and tin, 36, 44, 46, 78
Tongs, 41
Toolmakers, 9, 40
Torchbearers, 107
Traders, 131

guilds, 31
Trappers, 82
Traveling metalworkers, 42
Tree trimmers, 111
Trenton Blistered Steel, 50
Tribal peoples, 40-41
Troubleshooters, 111
Trustees, 108
Tudor, Mary, 115
Turquoise mines, 63
Twentieth century: see Modern
    times
Twigg, John, 138-139

Unions, 23
  communist nations, 24
  competition from nonunionized
    workers, 24-25
United Automobile Workers, 121
United Mine Workers, 95-96
United States
  Appalachian region, 96
  coal, 88-89
  mining, 87-88
  Oil Patch, 149
  petroleum and natural gas
    industry, 147-148
  steel, 51
  Virginia Armory, 137-138
  War of Independence, 117
  waterworks, 146-147
  weapon makers, 140, 142-143
  see also Colonial America
Upholsterers, 115
Uranium, 97
Urban guilds, 11-12
Urnfield peoples, 66
Utilities, 109

Vehicle makers, 113-122
  see also Automobile workers
    and automobiles; Coaches
    and carriages
Vespasian (emperor), 7-8
Vickers works, 139, 141
"Village Blacksmith, The", 55
Virginia Armory, 137-138
Vitruvius, 102
Vulcan, 128

Wagonmakers, 29
Wales, Prince of, 118
War Scare of 1930, 141-142
Warriors, 128-129, 134
Watchmen, 78
Water commissioner, 146
Waterwheels, 7, 40, 102
Way of the Warrior doctrine, 134
Wayland the Smith, 128

Weapon makers, 125-144
    Middle Ages, 129-130
Welders, 59
Well diggers and drillers, 145-150
Wendel, Jean Martin, 138
Whalers, 107
Wheels, 40
Wheelwrights, 113, 114, 115
White Monks: *see* Cistercian monks
White tawyers, 11
Whitney, Eli, 139
Wildcatters, 147
Williamson, George, 137
Winchester rifle, 139
Windlass, 84-85

Windmills, 102-105
Women
  coal mining and, 84-85
  mechanics and repairers, 33
  metalsmiths, 57
  textile industries, 17-18
Woodworkers, 40
Wootz steel, 38
Workers: *see specific types*, i.e.,
    Factory workers

Xenophon, 6

Zelevski, 142